Dominik Weishaupt · Victor D. Köchli · Borut Marincek

How Does MRI Work?

Dominik Weishaupt
Victor D. Köchli
Borut Marincek

How Does MRI Work?

An Introduction to the Physics and Function of Magnetic Resonance Imaging

Second Edition

Contributors:
J. M. Froehlich, D. Nanz, K. P. Pruessmann

With 57 Figures and 9 Tables

 Springer

Dominik Weishaupt, MD
Associate Professor
Institute of Diagnostic Radiology
Raemistrasse 100, 8091 Zurich, Switzerland

Victor D. Köchli, MD
Rötelstrasse 30, 8006 Zurich, Switzerland

Borut Marincek, MD
Professor and Chairman
Institute of Diagnostic Radiology
Raemistrasse 100, 8091 Zurich, Switzerland

Contributors:
Klaas P. Pruessmann, PhD
Assistant Professor
Institute of Biomedical Engineering
Swiss Federal Institute of Technology
Gloriastrasse 35, 8092 Zurich, Switzerland

Johannes M. Froehlich, PhD
Guerbet AG
Winterthurerstrasse 92, 8006 Zurich,
Switzerland

Daniel Nanz, PhD
Department of Medical Radiology
Raemistrasse 100, 8091 Zurich, Switzerland

Translator:
Bettina Herwig
Hauptstraße 4 H, 10317 Berlin, Germany

Corrected 2nd printing 2008

ISBN 978-3-540-30067-0   e-ISBN 978-3-540-37845-7

DOI 10.1007 / 978-3-540-37845-7

Library of Congress Control Number: 2006924129

Cover design: Frido Steinen-Broo, eStudio Calamar, Spain
Production & Typesetting: LE-Tex Jelonek, Schmidt & Vöckler GbR, Leipzig

Printed on acid-free paper

9 8 7 6 5 4 3 2 1

springer.com

Preface

It is with great pleasure that we present this completely revised English edition of our book *How Does MRI Work? An Introduction to the Physics and Function of Magnetic Resonance Imaging* only two years after publication of the first English edition. We are particularly pleased that our introductory textbook met with great approval in the English-speaking world and not just in the German-speaking countries. This success has been an enormous incentive for us to further improve and update the text. For this reason, we are now presenting a second edition. All chapters have been thoroughly revised and updated to include the latest developments in the ever-changing field of MRI technology. In particular, the chapter on cardiovascular imaging has been improved and expanded. We gratefully acknowledge the contribution of Daniel Nanz, PhD, the author of this chapter. Moreover, two completely new chapters have been added: "Fat Suppression Techniques" and "High-Field Clinical MR Imaging".

Notwithstanding these additions, the intended readership of our book remains the same: it is not a book for MR specialists or MR physicists but for our students, residents, and technologists, in short, all those who are interested in MRI and are looking for an easy-to-understand introduction to the technical basis of this imaging modality at the beginning of their MRI training.

The second English edition presented here corresponds to and appears together with the completely revised fifth German edition.

The authors gratefully acknowledge the support of numerous persons without whose contributions the new German and English editions of our book would not have been possible. First of all, we thank our readers, in particular those who bought and read the preceding versions and provided oral and written comments with valuable suggestions for improvement.

We should furthermore like to thank Klaas P. Pruessmann, PhD, and Johannes M. Froehlich, PhD, for their excellent introductions to parallel imaging and MR contrast agents.

Special thanks are due to our translator, Bettina Herwig, who very knowledgeably and with great care translated the entire text and provided valuable advice in preparing the new edition.

Finally, we would like to thank Springer-Verlag, in particular Dr. U. Heilmann, W. McHugh, and Dr. L. Ruettinger, for their cooperation.

For the authors:
Dominik Weishaupt, MD   January 2006

Contents

Abbreviations

FID	Free induction decay
FSE	Fast spin echo
GRE	Gradient echo
IR	Inversion recovery
MHz	Megahertz
MR	Magnetic resonance
MRA	Magnetic resonance angiography
MRI	Magnetic resonance imaging
msec	Milliseconds
NMR	Net magnetization vector
PC MRA	Phase-contrast MR angiography
PD	Proton density
ppm	Parts per million
RF	Radiofrequency
SAR	Specific absorption rate
SE	Spin echo
SNR	Signal-to-noise ratio
T	Tesla
TE	Echo time
TOF	Time of flight
TR	Repetition time

Note   In this book, the terms "z-direction" and "xy-plane" are frequently used. In all figures, the main magnetic field, B_0, is represented from bottom to top and its direction is designated by z. The other two dimensions of the magnetic field are denoted by x and y. The xy-plane is perpendicular to the z-axis and is thus represented horizontally in the figures.

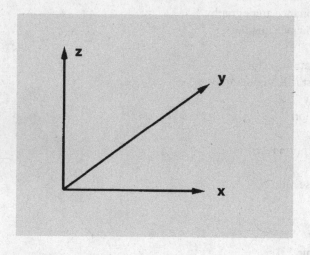

1 Spin and the Nuclear Magnetic Resonance Phenomenon

Medical magnetic resonance (MR) imaging uses the signal from the nuclei of hydrogen atoms (^{1}H) for image generation.

A hydrogen atom consists of a nucleus containing a single *proton* and of a single electron orbiting the nucleus (► Fig. 1). The proton having a positive charge and the electron a negative charge, the hydrogen atom as a whole is electrically neutral. The proton is of interest here.

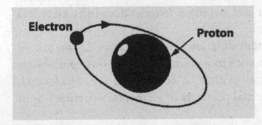

Electron **Proton**

Fig. 1.

Apart from its positive charge, the proton possesses *spin*, an intrinsic property of nearly all elementary particles. This means that the proton rotates about its axis like a spinning top. Such a proton has two important properties:

As a rotating mass (m), the proton has *angular momentum* and acts like a spinning top that strives to retain the spatial orientation of its rotation axis (► Fig. 2a).

As a rotating mass with an electrical charge, the proton additionally has *magnetic moment (B)* and behaves like a small magnet. Therefore, the pro-

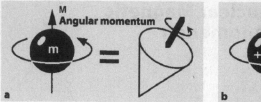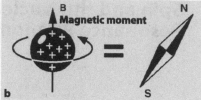

Fig. 2.

ton is affected by external magnetic fields and electromagnetic waves and, when it moves, induces a voltage in a receiver coil (▶ Fig. 2b).

A hydrogen nucleus differs from a spinning top, however, in that we cannot look into it and thus cannot see its intrinsic angular momentum, or spin, from the outside. In this respect, the nucleus is a black box for us. Nevertheless, we can identify the *orientation of its rotation axis* from the magnetization vector B. Thus, when we describe the rotation of a proton, we are not referring to its (invisible) angular momentum but to the "visible" motion of its magnetic axis, B. This motion is visible, so to speak, because it generates a signal in a receiver coil just like a magnet does in an electrical generator (e.g. a bicycle dynamo).

There is another, very important difference: while a spinning top can be slowed down and thus finally comes to a standstill, a proton's spin always has the same magnitude and can neither be accelerated nor decelerated, precisely because it is a fundamental property of elementary particles. Spin is simply there all the time!

How will a spin behave when brought into a strong magnetic field? To answer this question, let us again consider the spinning top:

When an external force (typically the earth's gravitational field G) acts on a spinning top and tries to alter the orientation of its rotational axis, the top begins to wobble, a process called *precession*. At the same time, friction at the point of contact withdraws energy from the spinning top and slows down its rotation. As a result, its axis becomes more and more inclined and the top finally falls over (▶ Fig. 3).

Once again, back to our hydrogen nuclei: when these are exposed to an external magnetic field, B_0, the magnetic moments, or spins, align with the direction of the field like compass needles. The magnetic moments do not only align with the field but, like spinning tops, undergo precession (▶ Fig. 4). Precession of the nuclei occurs at a characteristic speed that

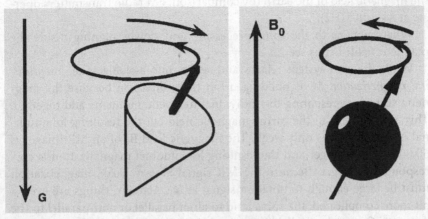

Fig. 3.   **Fig. 4.**

is proportional to the strength of the applied magnetic field and is called *Larmor frequency*. Alignment of the spins parallel to the magnetic field is a gradual process and, as with spinning tops, is associated with the dissipation of energy (► Chapter 2.1).

The Larmor frequency is a very important concept that is at the core of MR imaging. Let us therefore repeat:

The *Larmor or precession frequency* is the rate at which spins wobble when placed in a magnetic field.

The Larmor frequency is *directly proportional to the strength (B_0) of the magnetic field* and is given by the *Larmor equation*:

$\omega_0 = \gamma_0 \cdot B_0$
where
— ω_0 is the Larmor frequency in megahertz [MHz],
— γ_0 the gyromagnetic ratio, a constant specific to a particular nucleus, and
— B_0 the strength of the magnetic field in tesla [T].

Protons have a gyromagnetic ratio of $\gamma = 42.58$ MHz/T, resulting in a Larmor frequency of 63.9 MHz at 1.5 T as opposed to only about 1 kHz in the magnetic field of the earth (by comparison, FM radio transmitters operate at 88–108 MHz).

What happens to the spins precessing and slowly aligning inside the magnetic field? Let us see …

While the spin system relaxes and settles into a stable state, *longitudinal magnetization* M_z is building up in the z-direction because the magnetic vectors representing the individual magnetic moments add together. This also happens in the earth's magnetic field but the resulting longitudinal magnetization is only weak. The magnetic field B_0 of an MR imager is 60,000 times stronger and the resulting longitudinal magnetization is correspondingly larger. Because the MR signal is very weak, magnetization must be large enough to obtain a signal at all. Actually, things are even a bit more complicated: the spins tend to align parallel or anti-parallel to the magnetic field with parallel alignment being slightly preferred because it is equivalent to spins residing in a more favorable energy state. Hence, under steady-state conditions, a slightly larger fraction aligns parallel to the main magnetic field. It is this small difference that actually produces the measurable net magnetization M_z and is represented by the *net magnetization vector (NMV)*. Since the energy difference between the two orientations depends on the strength of the external magnetic field, M_z increases with the field strength.

Energy can be introduced into such a stable spin system by applying an electromagnetic wave of the same frequency as the Larmor frequency. This is called the *resonance condition*. The required electromagnetic wave is generated in a powerful radio transmitter and applied to the object to be imaged by means of an antenna coil. The process of energy absorption is known as excitation of the spin system and results in the longitudinal magnetization being more and more tipped away from the z-axis toward the transverse (xy-)plane perpendicular to the direction of the main magnetic field.

All of the longitudinal magnetization is rotated into the transverse plane by a radiofrequency (RF) pulse that is strong enough and applied long enough to tip the magnetization by exactly 90° (*90° RF pulse*). The resulting magnetization is now denoted by M_{xy} rather than M_z because it now lies in the xy-plane. Whenever transverse magnetization is present, it rotates or precesses about the z-axis, which has the effect of an electrical generator and induces an alternating voltage of the same frequency as the Larmor frequency in a receiver coil: the *MR signal*. This signal is collected and processed with sensitive receivers and computers to generate the MR image. The process of excitation of a spin system is illustrated graphically in ► Fig. 5.

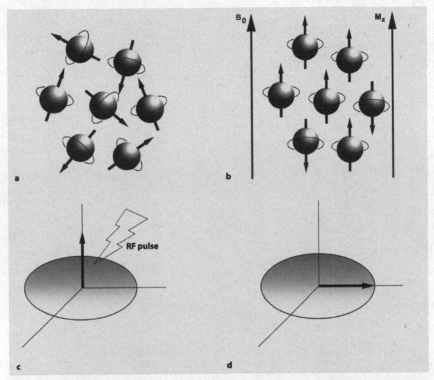

Fig. 5a–d. With no external magnetic field present, spins rotate about their axes in random direction (**a**). In the presence of a magnetic field, slightly more spins align parallel to the main magnetic field, B_0, and thus produce longitudinal magnetization, M_z (**b**). An RF pulse (**c**) tips the magnetization vector by exactly 90°, causing the entire longitudinal magnetization to flip over and rotate into transverse magnetization, M_{xy} (**d**)

References

1. Hobbie RK (1988) Magnetic resonance imaging. In: Intermediate physics for medicine and biology. Wiley & Sons, New York
2. Elster AD, Burdette JH (2001) Questions and answers in magnetic resonance imaging, 2nd edn. Mosby, St. Louis

2 Relaxation

What happens to the spins after they have been excited as just described?

Immediately after excitation, the magnetization rotates in the xy-plane and is now called *transverse magnetization or M$_{xy}$*. It is the rotating transverse magnetization that gives rise to the MR signal in the receiver coil. However, the MR signal rapidly fades due to two independent processes that reduce transverse magnetization and thus cause a return to the stable state present before excitation: spin-lattice interaction and spin-spin interaction. These two processes cause *T1 relaxation* and *T2 relaxation*, respectively.

2.1 T1: Longitudinal Relaxation

Transverse magnetization decays and the magnetic moments gradually re-align with the z-axis of the main magnetic field B$_0$, as discussed previously. The transverse magnetization remaining within the xy-plane – strictly speaking the projection of the magnetization vector onto the xy-plane (▶ Fig. 6) – decreases slowly and the MR signal fades in proportion. As transverse magnetization decays, the longitudinal magnetization, M$_z$ – the projection of the magnetization vector onto the z-axis – is slowly restored. This process is known as *longitudinal relaxation* or T1 recovery.

The nuclei can return to the ground state only by dissipating their excess energy to their surroundings (the "lattice", which is why this kind of relaxation is also called spin-lattice relaxation). The time constant for this recovery is *T1* and is dependent on the strength of the external magnetic field, B$_0$, and the internal motion of the molecules (Brownian motion). Biological tissues have T1 values of half a second to several seconds at 1.5 T.

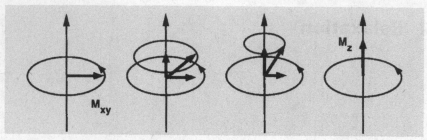

Fig. 6. T1 relaxation. Decay of transverse magnetization and regrowth of magnetization along the z-axis require an exchange of energy

2.2 T2/T2*: Transverse Relaxation

To understand transverse relaxation, it is first necessary to know what is meant by "phase". As used here, phase refers to the position of a magnetic moment on its circular precessional path and is expressed as an angle. Consider two spins, A and B, precessing at the same speed in the xy-plane. If B is ahead of A in its angular motion by 10°, then we can say that B has a phase of +10 relative to A. Conversely, a spin C that is behind A by 30° has a phase of –30° (▶ Fig. 7).

Immediately after excitation, part of the spins precess synchronously. These spins have a phase of 0° and are said to be in phase. This state is called *phase coherence*.

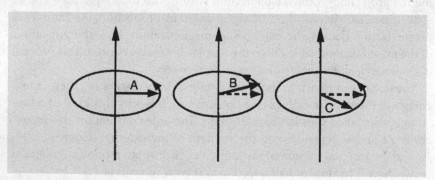

Fig. 7. Phase. Vector *B* has a phase of +10° relative to *A* while *C* has a phase of –30°. Note that all vectors rotate about the z-axis while their phases differ by the respective angles

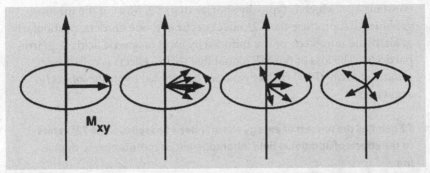

Fig. 8. T2 and T2* relaxation. Spins get out of phase (lose phase coherence), resulting in the loss of transverse magnetization without energy dissipation

For reasons that we will go into soon, phase coherence is gradually lost as some spins advance while others fall behind on their precessional paths. The individual magnetization vectors begin to cancel each other out instead of adding together. The resulting vector sum, the transverse magnetization, becomes smaller and smaller and finally disappears, and with it the MR signal (▶ Fig. 8).

In other words, transverse relaxation is the *decay of transverse magnetization because spins lose coherence (dephasing)*. Transverse relaxation differs from longitudinal relaxation in that the spins do not dissipate energy to their surroundings but instead exchange energy with each other. Coherence is lost in two ways:

— *Energy transfer* between spins as a result of *local changes in the magnetic field*. Such fluctuations are due to the fact that the spins are associated with small magnet fields that randomly interact with each other. Spins precess faster or slower according to the magnetic field variations they experience. The overall result is a cumulative loss of phase. It is a process due to *pure spin-spin interaction* and as such is unaffected by application of a 180° refocusing pulse (▶ Chapter 7). Dephasing occurs with the time constant T2 and is more or less independent of the strength of the external magnetic field, B_0.

— *Time-independent inhomogeneities of the external magnetic field B_0*. These are intrinsic inhomogeneities that are caused by the magnetic field generator itself and by the very person being imaged. They contribute to *dephasing*, resulting in an overall signal decay that is even faster than described by T2. This second type of decay occurs with the time constant T2*, which is typically shorter than T2. Most of the inhomo-geneities that produce the T2* effect occur at tissue borders, particularly at air/tissue interfaces, or are induced by local magnetic fields (e.g. iron particles). The loss of the MR signal due to T2* effects is called *free induction decay (FID)*. T2* effects can be avoided by using spin echo sequences.

T2 denotes the process of energy transfer between spins, while T2* refers to the effects of additional field inhomogeneities contributing to dephasing.

T1 and T2 relaxation are completely *independent* of each other but oc-cur more or less *simultaneously*! The decrease in the MR signal due to T2 relaxation occurs within the first 100–300 msec, which is long before there has been complete recovery of longitudinal magnetization M_z due to T1 re-laxation (0.5–5 sec).

References

1. Gore JC, Kennan RP (1999) Physical principles and physiological basis of magnetic relaxation. In: Stark DD, Bradley WG Jr (eds). Magnetic resonance imaging, 3rd ed. Mosby-Year Book no 33, Mosby, St. Louis
2. Elster AD, Burdette JH (2001) Questions and answers in magnetic resonance imaging, 2nd ed. Mosby, St. Louis
3. Elster AD (1986) Magnetic resonance imaging, a reference guide and atlas. Lippincott, Philadelphia

3 Image Contrast

What determines the contrast of an MR image and how can we influence it?

Having explained the concepts of excitation and relaxation, we can now answer this question. *Three intrinsic features* of a biological tissue contribute to its signal intensity or brightness on an MR image and hence image contrast:

— The *proton density*, i.e. the number of excitable spins per unit volume, determines the maximum signal that can be obtained from a given tissue. Proton density can be emphasized by minimizing the other two parameters, T1 and T2. Such images are called *proton density-weighted* or simply *proton density images*.

— The *T1 time* of a tissue is the time it takes for the excited spins to recover and be available for the next excitation. T1 affects signal intensity indirectly and can be varied at random. Images with contrast that is mainly determined by T1 are called *T1-weighted images (T1w)*.

— The *T2 time* mostly determines how quickly an MR signal fades after excitation. The T2 contrast of an MR image can be controlled by the operator as well. Images with contrast that is mainly determined by T2 are called *T2-weighted images (T2w)*.

Proton density and T1 and T2 times are intrinsic features of biological tissues and may vary widely from one tissue to the next. Depending on which of these parameters is emphasized in an MR sequence, the resulting images differ in their tissue-tissue contrast. This provides the basis for the exquisite soft-tissue discrimination and diagnostic potential of MR imaging: based on their specific differences in terms of these three parameters, tissues that are virtually indistinct on computed tomography (CT) scans can be differentiated by MRI without contrast medium administration.

3.1   Repetition Time (TR) and T1 Weighting

In order to generate an MR image, a slice must be excited and the resulting signal recorded many times. Why this is so will be explained in ▶ Chapter 4.

> *Repetition time (TR)* is the interval between two successive excitations of the same slice.

Repetition time (TR) is the length of the relaxation period between two excitation pulses and is therefore crucial for T1 contrast. When TR is long, more excited spins rotate back into the z-plane and contribute to the regrowth of longitudinal magnetization. The more longitudinal magnetization can be excited with the next RF pulse, the larger the MR signal that can be collected.

If a *short* repetition time (less than about 600 msec) is selected, image contrast is strongly affected by T1 (TR A in ▶ Fig. 9). Under this condition, tissues with a short T1 relax quickly and give a large signal after the next RF pulse (and hence appear bright on the image). Tissues with a long T1, on the other hand, undergo only little relaxation between two RF pulses and hence less longitudinal magnetization is available when the next excitation pulse is

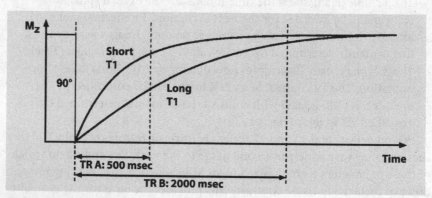

Fig. 9. Relationship between TR and T1 contrast. When TR is short (A), a tissue with a short T1 regains most of its longitudinal magnetization during the TR interval and hence produces a large MR signal after the next excitation pulse whereas a tissue with a long T1 gives only a small signal. When TR is long (B), the signal differences disappear because there is enough time for regrowth of longitudinal magnetization in both tissues

applied. These tissues therefore emit less signal than tissues with a short T1 and appear dark. An image acquired with a short TR is *T1-weighted* because it contains mostly T1 information.

If a fairly *long* repetition time (typically over 1500 msec) is selected, all tissues including those with a long T1 have enough time to return to equilibrium and hence they all give similar signals (TR B in ▸ Fig. 9). As a result, there is *less T1 weighting* because the effect of T1 on image contrast is only small.

Thus, by selecting the repetition time, we can control the degree of T1 weighting of the resulting MR image:

Short TR → strong T1 weighting
Long TR → low T1 weighting

The relationship between the MR signal of a tissue and its appearance on T1-weighted images is as follows:

Tissues with a *short T1* appear *bright* because they regain most of their longitudinal magnetization during the TR interval and thus produce a stronger MR signal.
Tissues with a *long T1* appear *dark* because they do not regain much of their longitudinal magnetization during the TR interval and thus produce a weaker MR signal.

3.2   Echo Time (TE) and T2 Weighting

What is an echo, anyway?

In ▸ Chapter 4 we will see that different gradients have to be applied to generate an MR image. For the time being it is sufficient to know that these gradients serve to induce controlled magnetic field inhomogeneities that are needed to encode the spatial origin of the MR signals. However, the gradients also contribute to spin dephasing. These effects must be reversed by applying a refocusing pulse before an adequate MR signal is obtained. The signal induced in the receiver coil after phase coherence has been restored is known as a *spin echo* and can be measured.

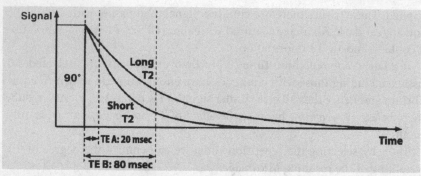

Fig. 10. Relationship between TE and T2 contrast. When TE is very short (A), there is virtually no signal difference between two tissues with different T2 times whereas clear differences become apparent when TE is longer (B): a tissue with a short T2 rapidly loses signal and becomes dark while a tissue with a long T2 retains its brighter signal for a longer time

> *Echo time (TE)* is the interval between application of the excitation pulse and collection of the MR signal.

The echo time determines the influence of T2 on image contrast. T2 is in the range of several hundred milliseconds and therefore much shorter than T1.

If a short echo time is used (less than about 30 msec), the signal differences between tissues are small (TE A in ► Fig. 10) because T2 relaxation has only just started and there has only been little signal decay at the time of echo collection. The resulting image has low T2 weighting.

If a longer echo time in the range of the T2 times of tissues (over about 60 msec) is used, the tissues are depicted with different signal intensities on the resulting MR image (TE B in ► Fig. 10): tissues with a short T2 having lost most of their signal appear dark on the image while tissues with a long T2 still produce a stronger signal and thus appear bright. This is why, for instance, cerebrospinal fluid (CSF) with its longer T2 (like water) is brighter on T2-weighted images compared with brain tissue.

By selecting an echo time (TE), the operator can control the degree of T2 weighting of the resulting MR image:

> Short TE → low T2 weighting
> Long TE → strong T2 weighting

▶ Fig. 10 also illustrates the relationship between the T2 value of a tissue and its appearance on T2-weighted images:

> **Tissues with a *short* T2 appear *dark* on T2-weighted images,**
> **tissues with a *long* T2 appear *bright* on T2-weighted images!**

The relationships between TR and TE and the resulting image contrast are summarized in ▶ Table 1. ▶ Table 2 lists the signal intensities of different tissues on T1- and T2-weighted images. ▶ Table 3 provides an overview of intrinsic contrast parameters of selected tissues.

A typical T1-weighted spin echo (SE) sequence is acquired with a TR/TE of 340/13 msec. A T2-weighted fast spin echo (FSE) MR image can be acquired with a TR/TE of 3500/120 msec. MR images that combine T1 and T2 effects are known as *proton density-weighted images (PD images)*. PD images with a TE of about 40 msec are also referred to as *intermediate-weighted images*. As a rule, PD images have a higher signal-to-noise ratio (▶ Chapter 5) than comparable T1- and T2-weighted images because the long TR allows recovery of longitudinal magnetization while the short TE minimizes the signal decrease due to the decay of transverse magnetization.

Typical parameters for acquisition of a PD image are for instance a TR/TE of 2000/15 msec for a PD-weighted SE sequence and a TR/TE of 4400/40 msec for a PD-weighted FSE sequence. PD sequences are especially useful for evaluating structures with low signal intensities such as the bones or connective tissue structures such as ligaments and tendons. Proton density weighting is often used for high-resolution imaging. SE sequences are preferred over FSE sequences for PD imaging because SE images are less prone to distortion. In the clinical setting, PD sequences are mainly used for imaging of the brain, spine, and musculoskeletal system.

3.3   Saturation at Short Repetition Times

In the section on repetition time, we already said that there is little time for the regrowth of longitudinal magnetization when TR is very short. The shorter the TR, the smaller the component of longitudinal magnetization that is restored and is available for subsequent excitation. As a consequence, the MR signal decreases as well. When a series of excitation pulses is applied, the MR signal becomes weaker and weaker after each repeat pulse. This process is known as *saturation* (▶ Fig. 11).

16

Table 1. Image contrast as a function of TR and TE

	TR	TE
T1-weighted	Short	Short
T2-weighted	Long	Long
Proton density-weighted (intermediate-weighted)	Long	Short

Table 2. Signal intensities of different tissues on T1- and T2-weighted images

Tissue	T1-weighted image	T2-weighted image
Fat	Bright	Bright
Aqueous liquid	Dark	Bright
Tumor	Dark	Bright
Inflammatory tissue	Dark	Bright
Muscle	Dark	Dark
Connective tissue	Dark	Dark
Hematoma, acute	Dark	Dark
Hematoma, subacute	Bright	Bright
Flowing blood	No signal due to black blood effect (▶ Chapter 7.2)	
Fibrous cartilage	Dark	Dark
Hyaline cartilage	Bright	Bright
Compact bone	Dark	Dark
Air	No signal	No signal

Table 3. Relative proton densities (%) and intrinsic T1 and T2 times (in msec) of different tissues

Tissue	Proton density	T1 (1.5 T)	T2 (1.5 T)
CSF	100	> 4000	> 2000
White matter	70	780	90
Gray matter	85	920	100
Meningioma	90	400	80
Metastasis	85	1800	85
Fat	100	260	80

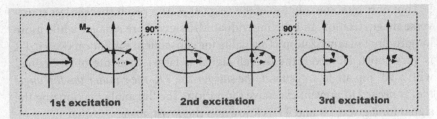

Fig. 11. Mechanism of saturation. With a very short TR, the longitudinal magnetization, M_z, that will recover in the interval and be available for subsequent excitation decreases after each RF pulse. In the example shown, the TR is so short that slightly less than half of the original longitudinal magnetization can regrow before the next excitation pulse is delivered

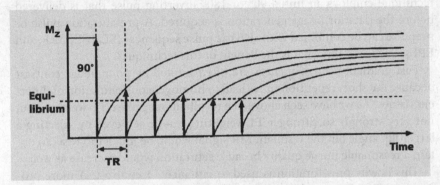

Fig. 12. Longitudinal magnetization at short repetition time. After repeat excitation at very short intervals, the amount of longitudinal magnetization, M_z, restored after each pulse settles at a low level (equilibrium or steady state). In this situation, the individual MR signals that form after each excitation are very weak

Saturation is an important issue when fast or ultrafast MR techniques are used. Here the MR signal may become very weak due to the very short repetition times (► Fig. 12). We will return to this phenomenon when we discuss gradient echo sequences.

3.4 Flip Angle (Tip Angle)

Partial flip angle imaging is a technique that can be used to minimize saturation and obtain an adequate MR signal despite a very short repetition time. A smaller flip angle does not deflect the magnetization all the way through

90° but only by some fraction of 90° (e.g. 30°). As a result there is less transverse magnetization and the individual MR signals are smaller while more longitudinal magnetization is available for subsequent excitation even if TR is very short. However, the overall signal is larger than the one obtained with a 90° flip angle. *In general, the shorter the TR, the smaller the flip angle that is needed* to prevent excessive saturation. The flip angle maximizing the signal for a given TR and TE is known as the *Ernst angle*.

3.5 Presaturation

Another option available to modulate image contrast is *presaturation*. This technique employs an initial 90° or 180° inverting pulse that is delivered before the data for image generation is acquired. A presaturation pulse or prepulse can be combined with all basic pulse sequences (SE, FSE, GRE, and EPI sequences). But what is the benefit of this technique?

Fast gradient echo sequences are often limited by poor image contrast because the short repetition times lead to homogeneous saturation of different tissues. As we have seen above, the resultant images are T1-weighted but not very strongly so. Stronger T1 weighting can be achieved by selecting a larger flip angle but the resultant MR signal would be much too weak to obtain a reasonable image quality because saturation would increase as well.

This is why presaturation is used to enhance T1 contrast. A more pronounced T1-effect is achieved with a 180° inverting pulse than with a 90° pulse because a 180° pulse inverts all longitudinal magnetization. As a result, T1 relaxation begins at −1 rather than 0 and twice as much longitudinal magnetization is available. Additionally, the operator can modulate the T1 effect by varying the time interval between the 180° inversion pulse and the excitation pulse (= inversion time, TI). TI can be chosen such that the signal contribution from a specific tissue is eliminated by applying the excitation pulse when the tissue has no magnetization. Thus, a short TI will suppress the signal from fat (▸ Chapter 7.5) and a long TI the signal from CSF (FLAIR sequence, ▸ Chapter 7.6). Another practical application is late-enhancement imaging in patients with myocardial infarction (▸ Chapter 11.8).

3.6   Magnetization Transfer

Without explicitly saying so, we have thus far always referred to free protons (i.e. protons in free water) when talking about protons because only these contribute to the MR signal. In addition to water protons, biological tissues also contain a specific pool of protons bound in macromolecules (usually proteins). These macromolecular protons cannot be directly visualized because of their very short T1. They have a wider range of Larmor frequencies than the water protons. This is why macromolecular protons can also be excited by RF pulses with frequencies slightly different from the Larmor frequency of hydrogen protons. Hence, it is possible to selectively excite a tissue with a large pool of macromolecular protons without directly affecting the protons in free water. Repeated delivery of the magnetization transfer pulse saturates the magnetization of the macromolecular protons from where it is transferred to free protons nearby. This process is associated with a drop in signal that depends in magnitude on the concentration of macromolecules and their interaction with free water and is known as *magnetization transfer* (▶ Fig. 13). The decrease in signal intensity due to magnetization transfer is large for solid tissues but only small for fluids (as long as their macromolecule content is low) and fatty tissue.

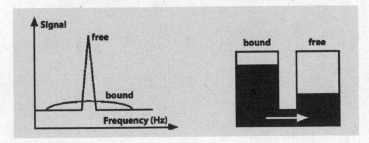

Fig. 13.

The phenomenon of magnetization transfer is exploited to improve image contrast using a technique known as *magnetization transfer imaging*. Magnetization transfer contrast (MTC) is used in cartilage imaging where it improves contrast between synovial fluid and cartilage because synovial fluid contains only few bound protons and thus shows only little magnetization transfer while cartilage contains a large proportion of bound protons and therefore shows pronounced magnetization transfer. In the brain, the MTC technique improves the detection of gadolinium-enhancing lesions.

References

1. Nessaiver M (1996) All you really need to know about MR imaging physics. University of Maryland Press, Baltimore
2. Duerk JL (1997) Relaxation and contrast in MR imaging. In: Riederer SJ, Wood ML (eds) Categorical course in physics: the basic physics of MR imaging. RSNA Publications no 19, Oak Brook
3. Elster AD, Burdette JH (2001) Questions and answers in magnetic resonance imaging, 2nd edn. Mosby, St. Louis

4 Slice Selection and Spatial Encoding

In the preceding sections, we have outlined the MR phenomenon and discussed the role of repetition and echo times. Now, finally, we want to make a picture! As a tomographic technique, MR imaging generates cross-sectional images of the human body. The excitation pulse is therefore delivered only to the slice we want to image and not to the whole body. How is this accomplished and how does the signal provide us with information about its origin within the slice?

For illustration, we consider a transverse (axial) slice or cross-section through the body. The magnetic field generated by most MR scanners is not directed from top to bottom, as in the illustrations we have used so far, but along the body axis of the person being imaged. From now on, this is the direction that will be designated by "z" since, as already said, *z stands for the direction of the main magnetic field*. The magnetic field gradients that now come into play are represented by wedges with the thick side indicating the higher field strength and the tip the lower field strength.

Both the excitation of a specific slice and the identification of the site of origin of a signal within the slice rely on the fact that the *precessional or Larmor frequency is proportional to the magnetic field strength*. In addition, recall that protons are excited only by an RF pulse with a frequency roughly equal to their Larmor frequency (*resonance condition*). If a uniform field of identical strength were generated throughout the body, all protons would have the same Larmor frequency and would be excited simultaneously by a single RF pulse.

To enable selective excitation of a desired slice, the magnetic field is therefore made *inhomogeneous* in a linear fashion along the z-direction by means of a gradient coil. As a result, the magnetic field strength has a smooth *gradient* so that, for example, it is weakest at the patient's head and strongest at the feet. The Larmor frequencies thus change gradually along the z-axis and

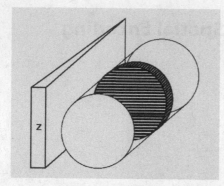

Fig. 14. Slice selection by means of the z-gradient. An RF pulse of a specific frequency excites exactly one slice (hatched) with adjacent slices being unaffected because they have different resonant frequencies

each slice now has its unique frequency. Hence, application of an RF pulse that matches the Larmor frequency of the desired slice will excite only protons within the chosen slice while the rest of the body remains unaffected (▶ Fig. 14).

Gradients are additional magnetic fields that are generated by gradient coils and add to or subtract from the main magnetic field. Depending on their position along the gradient, protons are temporarily exposed to magnetic fields of different strength and hence differ in their precessional frequencies. A shallow gradient generates a thicker slice while a steep gradient generates a thinner slice (▶ Fig. 15a). Slice position is defined by changing the center frequency of the RF pulse applied (▶ Fig. 15b).

Having selected slice position and thickness by application of an appropriate slice-select gradient, we can now proceed to explain how the spatial position of an MR signal is identified. This is accomplished by *spatial encoding*, which is the most difficult task in generating an MR image and requires the application of additional gradients that alter the magnetic field strength along the y- and x-axes. Once we have grasped the concept of spatial encoding, it will be easy to understand the different kinds of artifacts that degrade MR image quality in clinical practice. Spatial encoding comprises two steps, *phase encoding* and *frequency encoding*. These two steps are discussed in their appropriate order, which means that we must first turn to the more difficult technique of phase encoding.

For phase encoding, a gradient in the y-direction (from top to bottom) is switched on after the spins have been excited and precess in the xy-plane. Such a *phase-encoding gradient* alters the Larmor frequencies of the spins according to their location along the gradient. As a result, the excited spins higher up in the scanner experience a stronger magnetic field and thus gain

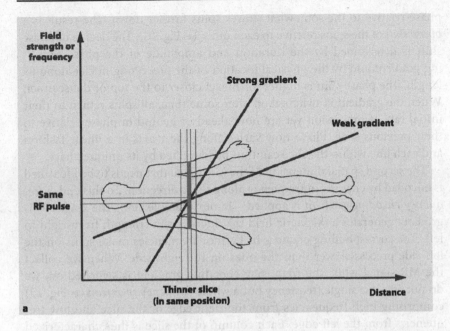

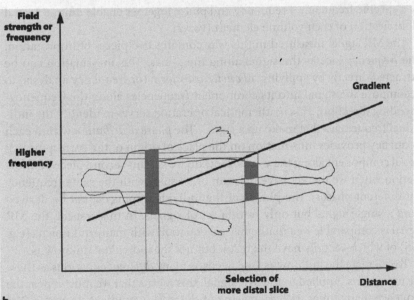

Fig. 15. a The strength of the gradient applied defines slice thickness. An RF pulse of a given frequency bandwidth produces a thin slice if the gradient is strong and a thick slice if the gradient is weak. **b** The center frequency of the RF pulse applied determines the location of the slice

phase relative to the somewhat slower spins further down. The result is a *phase shift* of the spins relative to each other (▶ Fig. 16). The degree of phase shift is determined by the duration and amplitude of the phase-encoding gradient and by the physical location of the precessing nuclei along its length. The phase gain is higher for nuclei closer to the top of the scanner. When the gradient is switched off after some time, all spins return to their initial rate of precession yet are now ahead or behind in phase relative to their previous state. Phase now varies along the y-axis in a linear fashion and each line within the slice can thus be identified by its unique phase.

The second spatial dimension of the MR signal that needs to be identified is encoded by changes in frequency along the x-direction. To this end, a *frequency-encoding gradient* is applied – in our example along the x-axis. This gradient generates a magnetic field that increases in strength from right to left. The corresponding changes in Larmor frequencies make spins on the left side precess slower than the ones on the right side. When we collect the MR signal while the frequency-encoding gradient is switched on, we do not obtain a single frequency but a whole *frequency spectrum* (▶ Fig. 17) comprising high frequencies from the right edge of the slice and low frequencies from the left edge. Each column of the slice is thus characterized by a specific frequency. Frequency and phase together enable unique spatial identification of each volume element (*voxel*).

The MR signal measured in this way contains two pieces of information. The *frequency* locates the signal along the x-axis. This information can be extracted directly by applying a *Fourier transform* (or frequency analysis) to decompose the signal into its component frequencies along the frequency-encoding direction. This mathematical operation serves to identify the individual frequencies that make up a signal. The *phase distribution* within each frequency provides information on the place of origin of the corresponding signal component along the y-axis. How do we get this second piece of information when we merely have the sum of all spins with the same frequency but different phases? The phases of the individual spins cannot be derived from a single signal but only from a set of signals. In this respect, the MR signal is comparable to a mathematical equation with many unknowns (e.g. 256) of which we only have the result but not the individual unknowns.

To calculate the unknowns, one needs as many *different* equations as there are unknowns. Applied to the MR signal, this means that we must repeat the sequence many times with increasing or decreasing gradient strengths. The set of echoes acquired with different phase encodings allows us to derive the required phase-encoded spatial information by applying a second Fourier

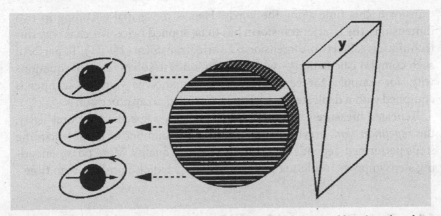

Fig. 16. Phase encoding by means of the y-gradient. Each horizontal line (e.g. the white line in the example) is identified by a unique amount of phase shift

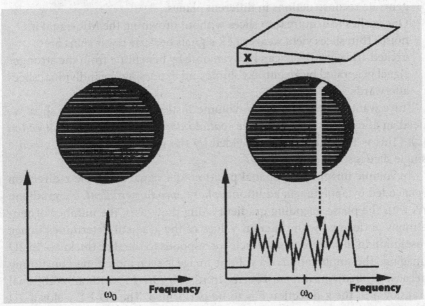

Fig. 17. Frequency encoding by means of the x-gradient. With the gradient switched off *(left)*, only a single frequency is received, the Larmor frequency ω_0. With the gradient switched on *(right)*, a frequency spectrum is received with each column being identified by its unique frequency

transform, this time along the y-axis. Hence, for spatial encoding in two dimensions, the Fourier transform has to be applied twice, which is why this technique is called two-dimensional Fourier transform (2D-FT). To perform such complex calculations – which corresponds to solving a set of equations with, for example, 256 equations and 256 unknowns – an MR scanner is equipped with a dedicated computer, a so-called array processor.

Repeated measurements are performed with a specific temporal delay, the *repetition time* previously mentioned. The number of phase-encoding steps performed depends on the desired image quality. More phase-encoding steps improve resolution and image quality but also prolong scan time.

4.1   Three-Dimensional Spatial Encoding

It is sometimes desirable to image a whole volume rather than just a number of individual slices, for the following reasons:
— The acquired source data set is to be postprocessed, for example, to generate reconstructions in different planes.
— One wishes to acquire thin slices without drowning the MR signal in noise. Thin slices yield weaker MR signals because fewer spins are excited. This drawback can be overcome by benefiting from the stronger signal generated by an entire volume and extracting the individual slices afterwards.

If we want to excite an entire volume instead of only a single slice, we need an additional step to encode *spatial information in the third direction (z)*. (This is the information provided by the slice-select gradient when a single slice is scanned.)

In volume imaging, the spatial position of a signal along the z-direction is encoded by applying an additional *phase-encoding gradient*, a z-gradient. As with the phase encoding gradient along the y-axis, the number of repetitions performed with different values of the gradient determines image resolution in the z-direction, which corresponds to the slice thickness in 2D imaging. The computation of a volume image is even more time-consuming because a *three-dimensional Fourier transform (3D-FT)* with an additional transform in the z-direction has to be performed. The 3D-FT yields a 3D data set of a volume without interslice gaps from which reconstructions in any plane or projections can be generated with the aid of suitable reconstruction algorithms. These techniques are very useful for MR angiography.

The major drawback of volume imaging is that it may unduly prolong scan time since spatial encoding in the x- and y-directions must be performed for each phase-encoding step along the z-axis.

4.2 K-Space

Data collected from the signals is stored in a mathematical area known as k-space. K-space has two axes with the horizontal axis (k_x) representing the frequency information and the vertical axis (k_y) the phase information (▶ Fig. 18). It is a graphic matrix of digitized MR data that represents the MR image before Fourier transformation is performed. Each line in k-space corresponds to one measurement and a line is acquired for each phase-encoding step. The center line (0) is filled with the data that is unaffected by the phase-encoding gradient (gradient isocenter).

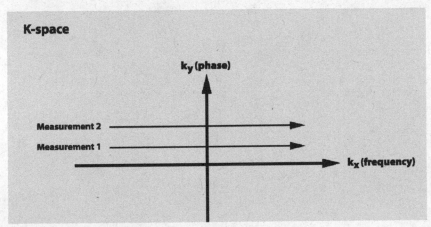

Fig. 18. K-space. k_x is the frequency axis, k_y the phase axis. The data from each measurement fills a different horizontal line

An MR image is created from the raw data by application of 2D-FT after the scan is over and k-space is filled. The lines in k-space *do not* correspond one to one with the lines in the resulting MR image. Rather, data in the *center of k-space* primarily determines *contrast* in the image while the *periphery (the outer lines)* primarily contains *spatial information*. When discussing fast sequences (▶ Chapter 8), we will also learn how we can speed up scanning by filling more than one k-space line with a single acquisition.

References

1. Wehrli FW (1997) Spatial encoding and k-space. In: Riederer SJ, Wood ML (eds). Categorical course in physics: The basic physics of MR imaging. RSNA Publications no 31, Oak Brook
2. Wood ML, Wehrli FW (1999) Principles of magnetic resonance imaging. In: Stark DD, Bradley WG Jr (eds) Magnetic resonance imaging, 3rd edn. Mosby-Year Book no 28, Mosby, St. Louis

5 Factors Affecting the Signal-to-Noise Ratio

In the preceding chapters we have learned how an MR signal is generated and how the collected signal is processed to create an MR image. What we have disregarded so far is that the MR signal can be degraded by noise. Image noise results from a number of different factors:

— Imperfections of the MR system such as magnetic field inhomogeneities, thermal noise from the RF coils, or nonlinearity of signal amplifiers.
— Factors associated with image processing itself.
— Patient-related factors resulting from body movement or respiratory motion.

The relationship between the MR signal and the amount of image noise present is expressed as the *signal-to-noise ratio (SNR)*. Mathematically, the SNR is the quotient of the signal intensity measured in a *region of interest (ROI)* and the standard deviation of the signal intensity in a region outside the anatomy or object being imaged (i.e. a region from which no tissue signal is obtained).

A high SNR is desirable in MRI. The SNR is dependent on the following parameters:

— Slice thickness and receiver bandwidth
— Field of view
— Size of the (image) matrix
— Number of acquisitions
— Scan parameters (TR, TE, flip angle)
— Magnetic field strength
— Selection of the transmit and receive coil (RF coil)

Before we discuss the effects of each of these parameters, it is first necessary to clarify some concepts.

5.1 Pixel, Voxel, Matrix

An MR image is digital and consists of a matrix of *pixels* or picture elements. A *matrix* is a two-dimensional grid of rows and columns. Each square of the grid is a pixel which is assigned a value that corresponds to a signal intensity. Each pixel of an MR image provides information on a corresponding three-dimensional volume element, termed a *voxel* (▶ Fig. 19). The voxel size determines the spatial resolution of an MR image.

The size of a voxel can be calculated from the field of view, the matrix size, and the slice thickness. In general, the resolution of an MR image increases as the voxel size decreases.

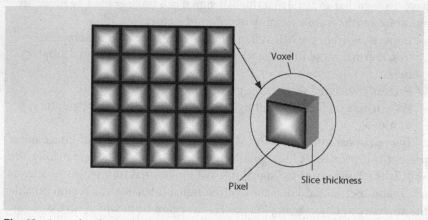

Fig. 19. A voxel is the tissue volume represented by a pixel in the two-dimensional MR image

5.2 Slice Thickness and Receiver Bandwidth

To achieve optimal image resolution, very thin slices with a high SNR are desirable. However, thinner slices are associated with more noise, and so the SNR decreases with the slice thickness. Conversely, thicker slices are associated with other problems such as an increase in partial volume effects.

The poorer SNR of thin slices can be compensated for to some extent by increasing the number of acquisitions or by a longer TR. Yet this is ac-

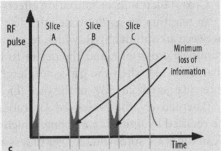

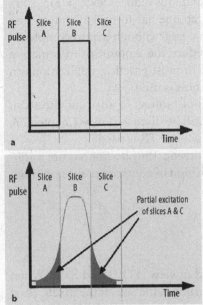

Fig. 20. **a** Ideal slice profile. **b** Distorted, non-rectangular slice profile in SE imaging with inadvertent excitation of adjacent slices reduces SNR. **c** With interslice gaps, the drop in SNR is minimized

complished only at the expense of the overall image acquisition time and reduces the cost efficiency of the MR imaging system.

The *receiver bandwidth* is the range of frequencies collected by an MR system during frequency encoding. The bandwidth is either set automatically or can be changed by the operator. A wide receiver bandwidth enables faster data acquisition and minimizes chemical shift artifacts (▶ Chapter 13.3) but also reduces SNR as more noise is included. Halving the bandwidth improves SNR by about 30%. With a narrow bandwidth, on the other hand, there will be more chemical shift and motion artifacts and the number of slices that can be acquired for a given TR is limited.

An *interslice gap* is a small space between two adjacent slices. It would be desirable to acquire contiguous slices but interslice gaps are necessary in SE imaging due to imperfections of RF pulses. Because the resultant slice profiles are not perfectly rectangular (▶ Fig. 20), two adjacent slices overlap at their edges when closely spaced. Under these conditions, the RF pulse for one slice also excites protons in adjacent slices. Such interference is known as *cross-talk*.

Cross-talk produces saturation effects and thus reduces SNR (► Fig. 20b).

In selecting an appropriate interslice gap one has to find a compromise between an optimal SNR, which requires a large enough gap to completely eliminate cross-talk, and the desire to reduce the amount of information that is missed when the gap is too large. In most practical applications an interslice gap of 25–50% of the slice thickness is used.

Alternatively, the undesired saturation of protons in adjacent slices can be reduced by *multislice imaging*, which will be discussed in ► Chapter 7.3. Scan times are somewhat longer unless a shorter TR is used.

Gradient echo (GRE) sequences are different. They do not require a 180° refocusing pulse and thus allow the acquisition of contiguous slices without interslice gaps.

5.3 Field of View and Matrix

There is a close relationship between field of view (FOV) and SNR. When matrix size is held constant, the FOV determines the size of the pixels. *Pixel size in the frequency-encoding direction* is calculated as the FOV in mm divided by the matrix in the frequency-encoding direction and *pixel size in the phase-encoding direction* as the FOV in mm divided by the matrix in the phase-encoding direction.

As illustrated in ► Fig. 21, pixel size changes with the FOV. A smaller FOV results in a smaller pixel size as long as the matrix is unchanged. Pixel size is crucial for the spatial resolution of the MR image. With the same FOV, a finer matrix (i.e. a matrix consisting of more pixels) results in an improved spatial resolution (► Figs. 22 and 23).

Conversely, a coarser matrix (i.e. one with fewer pixels) results in a poorer spatial resolution when the FOV is held constant (► Fig. 23).

From what has been said so far, one might conclude that the matrix should be as large as possible in order to encompass a maximum of picture elements. This is true in terms of image resolution but the minimum pixel size is limited by the fact that, in general, *SNR decreases with the size of the voxel*.

Another limiting factor is image acquisition or scan time, which increases in direct proportion to the matrix size. *Scan time* is the key to the economic efficiency of all MR systems and can be calculated by a simple equation.

Scan time = TR × number of phase-encoding steps × number of signal averages (NSA) [echo train length (ETL)].

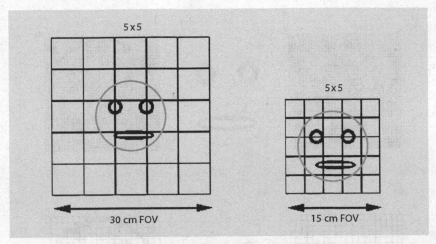

Fig. 21. Effect of the FOV on pixel size with the matrix size held constant

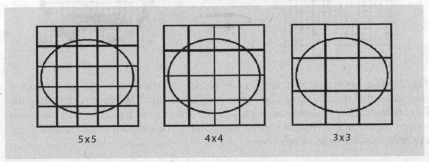

Fig. 22. A smaller matrix size with the FOV held constant results in larger pixels and thus a poorer spatial resolution

A "trick" can be used to achieve a high spatial resolution in a reasonable scan time. This is done by reducing the field of view only in the phase-encoding direction *(rectangular field of view)* and is possible because spatial resolution is determined by the matrix size in the frequency-encoding direction while scan time is determined by the matrix size in the phase-encoding direction. Reduction of the matrix size in the phase-encoding direction therefore does not reduce spatial resolution. Filling only one-half the normal number of phase-encoding lines in k-space reduces imaging time and the FOV by 50%. However, use of a rectangular FOV may be associated with wraparound artifacts when signal outside the FOV in the phase-encoding direction is mapped back into the image at an incorrect location

34

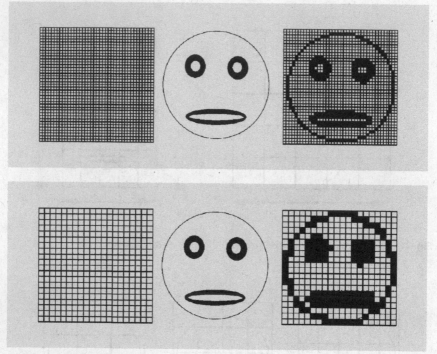

Fig. 23. Effect of matrix size on spatial resolution. Consider we are imaging a smiley face with a fine matrix (*top*) and a coarse matrix (*bottom*). The pixels representing the face are black. The two depictions of the face illustrate the much poorer detail resolution when a coarser matrix (*bottom right*) is used: pupil and eye cannot be distinguished and the open mouth appears to be closed

(▶ Chapter 13). This kind of foldover can be suppressed by specific anti-aliasing options such as "no phase wrap". Moreover, reduction of the FOV in the phase-encoding direction is associated with a slight drop in SNR. A rectangular FOV is typically used to image the spine and extremities and for MR angiography.

Scan time can be shortened further on state-of-the-art scanners that allow one to use rectangular fields of view in combination with rectangular pixels.

Finally, various *techniques of partial k-space acquisition* (▶ Figs. 24, 25, and 26) save scan time without one having to change the voxel size. In *partial Fourier imaging*, only half the lines (or slightly more) in the phase-encoding direction are filled (▶ Fig. 24) while *fractional* or *partial echo imaging* (▶ Fig. 25) refers to a technique with incomplete filling of the frequency-

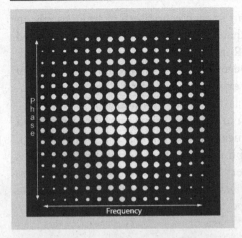

Fig. 24. Complete k-space sampling. Each data point represents one frequency-encoding line and one phase-encoding line

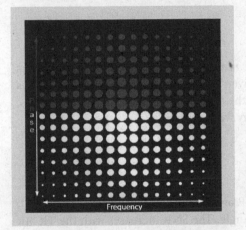

Fig. 25. Partial Fourier imaging. Slightly less than half the k-space lines in the phase-encoding direction are not sampled (*gray dots*). These lines are interpolated

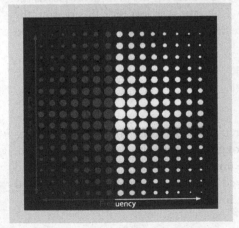

Fig. 26. Fractional echo imaging. Slightly less than half of the k-space lines in the frequency-encoding direction are not filled directly (*gray dots*). The unfilled lines represent the echo portions that have not been sampled. The resulting MR image has a similar resolution but poorer SNR compared with an image generated with complete k-space sampling (▶ Fig. 24) (as less "true" data is incorporated)

encoding lines by sampling only part of each echo. Both techniques rely on the inherent symmetry of k-space that allows one to interpolate the unfilled lines and to thus reconstruct an MR image when only half or slightly more than half the lines of k-space have been sampled. Both methods *shorten scan time* but this is accomplished *at the expense of SNR*. Partial Fourier and fractional echo imaging are needed for fast imaging techniques (▶ Chapter 8).

In routine 2D Fourier transform or spin-warp imaging, k-space is filled sequentially one line at a time (linear or Cartesian k-space acquisition). More sophisticated sequences use spiral k-space trajectories that fill the lines from the center toward the periphery (*elliptical centric ordering of k-space, CENTRA*). In MR angiography, for instance, this technique is used to fill the center of k-space with the data important for evaluating contrast enhancement patterns.

5.4 Number of Excitations

The *number of excitations (NEX)* or *number of signal averages (NSA)* denotes how many times a signal from a given slice is measured. The SNR, which is proportional to the square root of the NEX, improves as the NEX increases, but scan time also increases linearly with the NEX.

5.5 Imaging Parameters

Other parameters affecting the SNR are the sequence used, echo time (TE), repetition time (TR), and the flip angle. The SNR increases with the TR but the T1 effect is also lost at longer TRs. Conversely, the SNR decreases as the TE increases. With a short TE, the T2 contrast is lost. For this reason, the option of shortening TE to improve SNR is available only for T1-weighted sequences.

5.6 Magnetic Field Strength

Applying a *higher magnetic field strength increases longitudinal magnetization* because more protons align along the main axis of the magnetic field, resulting in an increase in SNR. The improved SNR achieved with high-field systems (▶ Chapter 14) can be utilized to generate images with an improved spatial resolution or to perform fast imaging.

5.7   Coils

An effective means to improve SNR, without increasing voxel size or length-ening scan time, is selecting an appropriate *radiofrequency (RF) coil*. In gen-eral, an RF coil should be as close as possible to the anatomy being imaged and surround the target organ. The nearer the coil can be placed to the or-gan under examination, the better the resulting signal. RF coils can be used either to transmit RF and receive the MR signal or to act as receiver coils only. In the latter case, excitation pulses are delivered by the body coil. The basic coil types that are distinguished are briefly described below.

5.7.1   Volume Coils

Volume coils may be used exclusively as *receive coils* or as *combined trans-mit/receive coils*. Volume coils completely surround the anatomy to be im-aged. Two widely used volume coil configurations are the *saddle coil* and the *birdcage coil*. Volume coils are characterized by a homogeneous signal qual-ity. Another type of volume coil is the *body coil*, which is an integral part of an MR scanner and is usually located within the bore of the magnet itself. Head and extremity coils are further examples of volume coils.

5.7.2   Surface Coils

Most surface coils can only receive the MR signal and rely on the body coil for delivery of RF pulses. Combined transmit/receive surface coils are also available. Surface coils are used for spinal MRI and imaging of small ana-tomic structures.

5.7.3   Intracavity Coils

Intracavity coils are small *local receive coils* that are inserted into body cavi-ties to improve image quality as a result of the closer vicinity to the target organ. In clinical MRI, endorectal coils are used for imaging of the prostate and the anal sphincter muscle. Experimental applications include endovas-cular imaging and imaging of hollow organs.

5.7.4   Phased-Array Coils

Phased-array coils serve to *receive* MR signals. A phased-array system consists of several independent coils connected in parallel or series. Each coil feeds into a separate receiver. The information from the individual receivers is combined to create one image. Phased-array coils yield images with a high spatial resolution and allow imaging with a larger field of view as they improve both SNR and signal homogeneity.

► Table 4 summarizes the factors affecting SNR.

► Table 5 summarizes the effects of matrix size, slice thickness, and FOV on spatial resolution.

► Table 6 summarizes the effects of different sequence parameters on scan time.

Table 4. Effects of different imaging and sequence parameters on signal-to-noise ratio (SNR)

Change in parameter	SNR
Increasing slice thickness	Increases
Increasing FOV	Increases
Reducing FOV in phase-encoding direction (rectangular FOV)	Decreases
Increasing TR	Increases
Increasing TE	Decreases
Increasing matrix size in frequency-encoding direction	Decreases
Increasing matrix size in phase-encoding direction	Decreases
Increasing NEX	Increases
Increasing magnetic field strength	Increases
Increasing receiver bandwidth	Decreases
Employing local coils	Increases
Partial Fourier imaging	Decreases
Fractional echo imaging	Decreases

Table 5. Effects of matrix size, slice thickness, and field of view (FOV) on spatial resolution

Change in parameter	Spatial resolution
Increasing matrix size	Increases
Using thicker slices	Decreases
Increasing FOV	Decreases

Table 6. Effects of different sequence parameters on scan time

Change in parameter	Scan time
Using thicker slices	Decreases
Increasing FOV	No direct effect
Using rectangular FOV (in phase-encoding direction)	Decreases
Increasing TR	Increases
Increasing TE	Increases
Increasing matrix size in frequency-encoding direction	Increases
Partial Fourier imaging	Decreases
Fractional echo imaging	Decreases
Increasing NEX	Increases

References

1. Elster AD, Burdette JH (2001) Questions and answers in magnetic resonance imaging, 2nd ed. Mosby, St. Louis
2. Mitchell DG, Cohen MS (2004) MRI principles, 2nd ed. Saunders, Philadelphia
3. Hendrick RE (1999) Image contrast and noise. In: Stark DD, Bradley WG Jr (eds) Magnetic resonance imaging, 3rd ed. Mosby-Year Book no 43. Mosby, St. Louis

6 The MR Scanner

All major components of an MRI system have now been mentioned. They are (▶ Fig. 27):
— A strong *magnet* to generate the static magnetic field (B_0).
— A *gradient system* consisting of three coils to produce linear field distortions in the x-, y-, and z-directions and the corresponding *amplifiers*.
— A *radiofrequency (RF) transmitter* with a transmit coil built into the scanner.
— A highly sensitive *RF receiver* to pick up and amplify the MR signal. Alternatively, imagers may use a single RF coil switched between the transmit and receive modes.
— Additional coils, either receive coils or transmit/receive coils.
— Various *computers* for controlling the scanner and the gradients (*control computer*), for creation of the MR images *(array processor)*, and for coordinating all processes (*main or host computer*, to which are connected the *operator's console* and *image archives*).
— Other *peripheral devices* such as a control for the patient table, electrocardiography (ECG) equipment and respiration monitors to trigger specialized MR sequences, a cooling system for the magnet, a second operator's console (e.g. for image processing), a device for film exposure, or a PACS (picture archiving and communications system).

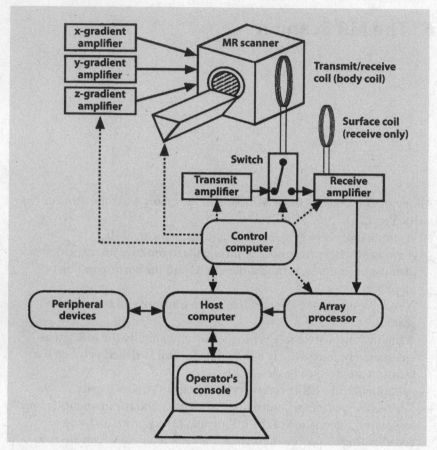

Fig. 27. The major components of an MR scanner

6.1 The Magnet

The main magnetic field generated by the magnet must have the following features:
— An adequate strength, which typically ranges from 0.1 to 3.0 T in medical MR imaging.
— A high stability without fluctuations in field strength.
— The best homogeneity possible with a uniform strength throughout the entire field and without "holes". Field homogeneity is usually expressed in ppm relative to the main field over a certain distance. Inhomogeneities throughout the scan volume should be below 5 ppm (0.0005%).

Three types of magnets are distinguished:
— *Resistive magnets* are conventional *electromagnets* that depend on a high
  and constant power supply to create a magnetic field. The maximum
  field strength generated by resistive magnets is about 0.3 T. Their major
  disadvantages are the high operating costs due to the large amounts of
  power required and a field homogeneity that is often poor. An advantage is the safety of the system as the field can be turned off instantly in
  an emergency.
— *Permanent magnets* consist of ferromagnetic substances and create a
  magnetic field that is maintained without an external power supply.
  However, permanent magnets are very heavy, can generate a field with
  a maximum strength of only 0.5 T, and rely on a constant external
  temperature.
— *Superconducting magnets* consist of a coil made of a *niobium-titanium
  (Nb-Ti) alloy* whose resistance to current flow is virtually eliminated
  when cooled to near absolute zero (about 4°Kelvin or −269°C). In this
  superconducting state, which is achieved using coolants known as
  cryogens (usually liquid helium), a current once induced flows practi-
  cally forever. Once the magnetic field has been established, it is maintained without additional power input. Very strong and highly homogeneous magnetic fields of up to 18 T can be generated using superconducting magnets. However, liquid helium evaporates and must be
  resupplied regularly. In an emergency it is not possible to simply switch
  off the magnet. About 95% of all MR systems used today have superconducting magnets. A *quench* refers to a magnet's sudden loss of
  superconductivity with subsequent breakdown of the magnetic field and
  may be induced by very minute movements of the coil. Due to the
  frictional energy released by this process, the coil temperature rises
  above the superconductivity threshold and the coils suddenly develop
  resistance. The current passing through an area of elevated coil resistance creates heat, which causes a sudden boiloff of cryogens. The
  risk of quenches is reduced by insulation of the Nb-Ti with an extra
  copper winding. Magnetic quenches are serious events but have become
  rare with the use of state-of-the-art magnet technology.

Magnetic field homogeneity is a primary consideration in medical MRI,
regardless of the magnet used. To achieve an optimal homogeneity, it is often
necessary to make adjustments known as *shimming*. This is done either pas-
sively by placing pieces of sheet metal at certain locations within the magnet
bore and on the outer surface of the scanner or actively by the activation of
specialized coils of which over 20 may be present in a scanner.

Another important aspect is shielding of the magnet, which serves to control the fringe fields external to the magnet. In the past, fringe fields were contained mainly by incorporating large amounts of iron into the walls and the ceiling of the scanner room (10–20 tons!). Because of weight and expense, this form of shielding is increasingly being abandoned and magnets with integrated or *active shielding* are used instead. Actively shielded magnets have a double set of windings of which the inner one creates the field while the outer one provides return paths for the magnetic field lines.

6.2 The Gradient System

Magnetic field gradients are applied for slice selection and spatial encoding (▶ Chapter 4). A set of three separate gradient coils, each with its own amplifier, is needed to alter the magnetic field strength along the x-, y-, and z-axes. These are switched on separately or in combination, e.g. to define an oblique slice. The *isocenter* is the geometric center of the main magnetic field, where the field strength is not affected by any of the three gradients. The gradient coils generate magnetic fields that are small compared with the main field but nevertheless need a current of several hundred amperes. The changing magnetic fields generated when the gradients are switched lead to the typical banging sound heard during an MR scan. Similar to a loudspeaker, which is nothing but a coil inside a magnetic field, the gradient coils "try to move" when the current is switched on and off, which causes a noisy clanging.

Despite the high currents, the gradient fields must be extremely stable in order to prevent image distortions. Moreover, it has been shown for gradient coils as well that actively shielded coils (▶ Chapter 6.1) are superior to the simpler versions: with smaller fringe fields, there is less external RF interference (induction of so-called eddy currents, ▶ Chapter 13.7).

Gradient performance is measured by three parameters:
— Maximum gradient strength (in units of mT/m)
— Rise time – time to maximum gradient amplitude
— Slew rate – maximum gradient amplitude/rise time

6.3   The Radiofrequency System

The *radiofrequency (RF) system* comprises a powerful *RF generator* (the Larmor frequency at 1.5 T is 63.8 MHz, which is in the range of FM transmitters) and a highly sensitive *receiver*. The stability of these two components is crucial: as both the frequency and the phase of the signal are needed for spatial encoding, any distortions, e.g. by phase rotation introduced by the receiver, would result in a blurred image. Moreover, to adequately detect the weak MR signal, effective RF shielding of the scanner room is necessary to prevent interference from external sources. This can be achieved by housing the magnet in a closed conductive structure known as a Faraday cage.

The RF subsystem also includes the transmit and receive coils. These may be combined coils acting as both transmitters and receivers such as the *body coil* which is integrated into the scanner. It is not visible from the outside and consists of a "cage" of copper windings encircling the patient. The RF transmitter serves to deliver pulses that correspond to the resonant frequency of hydrogen atoms.

As discussed in ▶ Chapter 5, the SNR can be modulated by employing coils other than the body coil. Careful coil selection according to the anatomy being imaged is important for optimizing image quality.

6.4   The Computer System

The computers of an MRI system control and coordinate many processes ranging from turning on and off gradients and the RF coils to data handling and image processing.

References

1.   McFall JR (1997) Hardware and coils for MR imaging. In: Riederer SJ, Wood ML (eds) Categorical course in physics: The basic physics of MR imaging. RSNA Publications no 41, Oak Brook

7 Basic Pulse Sequences

Let us once again go through the different steps that make up an *MR pulse sequence*.

— Excitation of the target area
  - Switching on the slice-selection gradient,
  - Delivering the excitation pulse (RF pulse),
  - Switching off the slice-selection gradient.
— Phase encoding
  - Switching on the *phase-encoding gradient* repeatedly, each time with a different strength, to create the desired number of phase shifts across the image.
— Formation of the echo or MR signal
  - Generating an echo, which can be done in two ways (discussed below).
— Collection of the signal
  - Switching on the *frequency-encoding or readout gradient*,
  - *Recording* the echo.

These steps are repeated many times, depending on the desired image quality. A wide variety of sequences are used in medical MR imaging. The most important ones are the spin echo (SE) sequence, the inversion recovery (IR) sequence, and the gradient echo (GRE) sequence, which are the basic MR pulse sequences.

We have already briefly mentioned *echoes* (► Chapter 3) and said that some time must elapse before an MR signal forms after the hydrogen protons have been excited. Now we can explain why this is so:

— Before an MR signal can be collected, the phase-encoding gradient must be switched on for spatial encoding of the signal.
— Some time is also needed to switch off the slice-selection gradient and switch on the frequency-encoding gradient.

— Finally, formation of the echo itself also takes time, which varies with the pulse sequence used.

7.1   Spin Echo (SE) Sequences

Spin echo sequences use a *slice-selective 90° RF pulse* for excitation, after which transverse magnetization decays with T2*, as discussed in ▸ Chapter 2. Dephasing occurs because some spins precess faster than others as a result of the static magnetic field inhomogeneities that are always present. This is why after half of the echo time (TE) has elapsed, a *180° RF pulse* is delivered to reverse or refocus the spins: those spins that were ahead before are now behind and vice versa. However, the spins that are now behind will catch up as they are still exposed to the same field inhomogeneities that caused the phase differences in the first place. Thus, after the second half of the TE interval has passed, all spins meet once again in phase. This is the moment at which the echo forms (▸ Fig. 28). The role of the 180° refocusing pulse in generating the spin echoes can be illustrated by considering a race in which a number of runners start together and, after some time, are given a signal to go back. At the time the signal is given, the fastest runners will have covered the longest distance but also have the longest way back. Assuming that everyone is still running *at their initial speed*, they will all arrive at the starting line together. (The analogy is not quite correct since it is not the direction of precession that is reversed but merely the position of the spins on the precessional path relative to each other. Applied to the example of the race, a magician would have to reverse the order of the runners without their noticing!)

The 180° refocusing pulse then serves to eliminate the effects of static magnetic field inhomogeneities (T2*) but cannot compensate for *variable* field inhomogeneities that underlie spin-spin interaction (T2). Therefore, the magnetization decay that occurs after excitation is slower as it is a function of T2 rather than T2*. Because of this decay, the transverse magnetization component is smaller at the time the echo is collected than immediately after excitation though the decrease in signal is less pronounced than it would be without application of the 180° refocusing pulse. Again, in our analogy, this means that not all runners arrive at the starting line together because they do not always run at a constant speed.

Spin echo sequences are characterized by an excellent image quality precisely because the effects of static field inhomogeneities are eliminated by

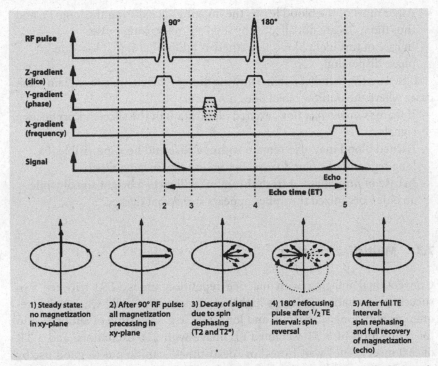

Fig. 28. SE sequence. The excitation pulse always has a flip angle of 90°; the dephased spins are refocused into the spin echo by the 180° pulse. The dashed lines indicate the phase-encoding steps

application of the 180° refocusing pulse. The tradeoff is a fairly long scan time, which makes the sequence highly sensitive to motion artifacts. SE sequences are still used as the standard sequences for acquiring T1-weighted or PD-weighted images. They are preferred for PD imaging because they are less susceptible to motion artifacts compared with FSE sequences.

7.2 Black Blood Effect

The *black blood effect*, or *outflow effect*, refers to a natural high contrast between flowing blood and tissue. It is a specific feature of SE sequences due to the long echo time. Flowing blood appears black because it does not give a signal. This has two reasons:

— All or most of the blood leaves the imaging slice during the long TE and thus the spins are not affected by the 180° refocusing pulse.
— In case of turbulent blood flow, there is additional signal loss due to phase dispersion.

Based on the fact that normal flowing blood is black, we can explain those cases where the outflow effect does not occur:

— If there is *slow blood flow*, excited blood stays in the slice and produces a signal.
— Excited blood may also remain within a slice and become visible if a long segment of a blood vessel lies *within the imaging slice*.
— In case of *thrombosis*, a fresh thrombus will yield a bright signal while an older, organized thrombus appears somewhat darker.

7.3   Multislice Imaging

Conventional imaging with inactive repetition times (TR) between two successive excitation pulses is highly inefficient, especially when using sequences with long scan times and long TRs (e.g. scan time of almost 3 min for acquisition of a T1-weighted SE image with 256 excitations and a TR of 500 msec). The "wait times" or "dead times" can be put to good use by exciting and recording signals from other slices during this period. In this way, 12 slices instead of only one can be acquired in the same time (or even up to 30 slices for T2-weighted sequences with TRs of 2000–4000 msec; ▶ Fig. 29).

A disadvantage of multislice imaging is that, due to imperfect slice profiles or RF pulses, protons outside the selected slice will also be excited. As a result, there will be less longitudinal magnetization and a weaker MR signal.

7.4   Inversion Recovery (IR) Sequences

Inversion recovery (IR) sequences are typically used for *T1-weighted or fat-suppressed imaging* but they can also be used to acquire T2-weighted images.

An IR sequence is an SE sequence with an additional 180° inversion pulse that precedes the usual 90° excitation pulse and 180° rephasing pulse of a conventional SE sequence. The inversion pulse flips longitudinal magnetiza-

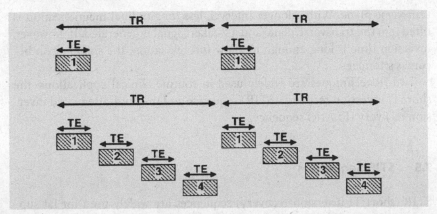

Fig. 29. Multislice imaging (interleaved acquisition). The inactive repetition time, TR, for the first slice is used productively to acquire data from other slices. In the example shown, we thus obtain four slices instead of only one in the same time. (The rectangles represent the different slices)

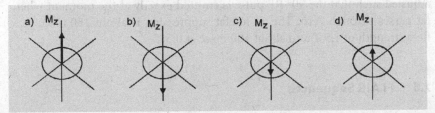

Fig. 30a–c. Inversion recovery sequence with T1 relaxation. Following the 180° inversion pulse (**a**), the longitudinal magnetization vector points in the opposite direction (**b**). T1 relaxation takes place from −z to +z (**c, d**). No signal forms as long as there is no vector component in the transverse plane (the null point of a tissue)

tion from the positive z-direction into the negative z-direction (► Fig. 30), which is indicated by the longitudinal magnetization vector now pointing in the opposite direction. As no component of the magnetization vector is in the transverse plane, no signal forms after delivery of the 180° RF pulse. Instead, the inverted longitudinal magnetization vector moves through the transverse plane to return to its original orientation. After some relaxation has occurred, the 90° pulse of the SE sequence is applied. The time between the 180° pulse and the 90° RF pulse is the *inversion time (TI)*.

Image contrast can be manipulated by changing the inversion time. With a short TI and delivery of the 90° excitation pulse immediately after the 180° inversion pulse, all negative longitudinal magnetization is flipped into the transverse plane. With a longer interval, less longitudinal magnetization is tilted into the transverse plane and a weaker signal is generated. If, however, inversion time is long enough to allow full relaxation, the signal again becomes stronger.

Two IR techniques are widely used in routine clinical applications: the short TI inversion recovery (STIR) sequence and the fluid-attenuated inversion recovery (FLAIR) sequence.

7.5   STIR Sequences

STIR (short TI inversion recovery) sequences are widely used for fat suppression because they reliably eliminate the signal from fat at all magnetic field strengths. A standard STIR sequence inverts the longitudinal magnetization of both fat and water by delivery of the 180° pulse, which is followed by a TI of some hundred milliseconds. To suppress the fat signal, the TI is adjusted such that the 90° RF pulse is emitted exactly at the moment when fat passes through zero. The TI for fat suppression is about 150 msec at a field strength of 1.5 T and about 100 msec at 0.5 T.

7.6   FLAIR Sequences

FLAIR (fluid-attenuated inversion recovery) is an inversion recovery technique that differs from STIR in that very long TI values (typically about 2000 msec) are used. Another difference is that FLAIR sequences are FSE sequences. With such long inversion times, there is nearly complete suppression of the signal from cerebrospinal fluid (CSF) while there is excellent detection of signals from brain tissue, tumors, edema, and fat. FLAIR sequences a very useful for detecting lesions with a poor contrast to surrounding brain tissue.

7.7   Gradient Echo (GRE) Sequences

Gradient echo sequences are also known as *gradient-recalled echo* or *fast-field echo* (FFE) sequences. As suggested by the name, GRE sequences employ *the gradient coils* for producing an echo rather than pairs of RF pulses. This is done by first applying a frequency-encoding gradient with negative polarity to destroy the phase coherence of the precessing spins (*dephasing*). Subsequently, the gradient is reversed and the spins *rephase* to form a gradient echo (► Fig. 31).

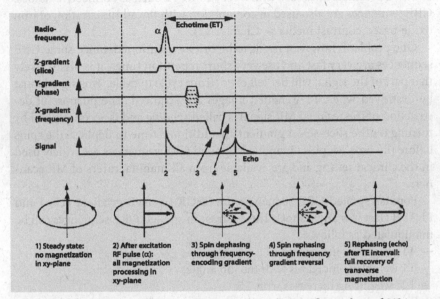

Fig. 31. Gradient echo sequence. For the sake of simplicity, a flip angle α of 90° is assumed here as well

Since no 180° refocusing pulse is needed to generate gradient echoes, very short repetition times (TR) can be achieved. As TR is a major determinant of the overall scan time of a GRE sequence – and of most other sequences – much *faster imaging* is possible compared with SE and IR sequences, which is the most important advantage of GRE imaging. As a result, GRE sequences are less frequently troubled by motion artifacts and are thus preferred whenever a short scan time is desirable. A disadvantage of a short TR is that the time available for T1 relaxation is also short. This may lead to saturation and reduce the SNR when a large flip angle is used. Because no 180° RF pulse is delivered, static field inhomogeneities are not compensated for and the signal decays with T2*. The image contrast resulting from differences in the T2* decay of various tissues is called *T2* contrast*. The T2* contrast of GRE images is affected by TE, which should be as short as possible to achieve optimal T1 weighting (to minimize T2* contrast and to reduce susceptibility effects). Conversely, a longer TE is selected to accentuate T2* contrast. T1 effects are minimized by simultaneously using a long TR. T2*-weighted images are useful to detect calcifications or deposits of blood products in tissues with a very short T2 such as connective tissues. GRE sequences are also used in conjunction with the administration of iron oxide-based contrast media (▶ Chapter 12).

One problem, however, needs to be briefly mentioned. Since some GRE sequences are very fast and use very short repetition times, it is highly likely that part of the signal will be "left over" from cycle to cycle. This signal must be destroyed when T1-weighted images are acquired. The purposeful destruction of the residual MR signal is called *spoiling* and is accomplished by turning on the slice-select gradient an additional time to dephase the spins before the next RF pulse is applied. Spoiled GRE sequences are widely used in the clinical setting and are available from all manufacturers of MR scanners.

Popular spoiled GRE sequences are SPGR (spoiled gradient echo) and FLASH (fast low angle shot). The contrast in spoiled GRE sequences can be manipulated as follows:
— T1 weighting increases as TR decreases;
— T1 weighting increases with the flip angle;
— T2* weighting increases with TE.

Proton density-weighted images are generated with a fairly long TR (100–400 msec), a low flip angle (≤20°), and a short TE (5–10 msec). T2*-weighted images result when a long TR (20–500 msec) and long TE (2–50 msec) are used. T1 weighting is achieved by a short TR (20–80 ms), short TE (5–10 msec), and a flip angle of 30–50°.

Spoiled GRE sequences can be acquired in the 2D or 3D mode. The 3D spoiled GRE technique enables volumetric thin-slice imaging without interslice gaps and allows for multiplanar reformatting.

A special type of GRE sequence used for routine MR imaging is the steady-state free precession (SSFP) sequence. SSFP is an unspoiled sequence in that part of the phase coherence of transverse magnetization is preserved from one TR interval to the next. This means that the transverse magnetization generated with a single RF pulse contributes to the formation of several echoes. Various acronyms are used by different manufacturers to designate SSFP sequences such as GRASS (gradient-recalled acquisition in the steady state) or FISP (fast imaging with steady-state precession). Further developments of the SSFP technique are FIESTA (fast imaging employing steady-state acquisition), balanced FFE (fast-field echo), and true FISP. FIESTA and true FISP are T2-weighted GRE sequences whose image contrast is determined by the T2/T1 ratio. Blood has a high T2/T1 ratio and therefore appears bright on SSFP images. Another advantage of SSFP is that it is not very prone to flowing blood. SSFP sequences are characterized by very short scan times and are thus well suited for vascular imaging and real-time imaging of moving organs such as the heart (▶ Chapter 11.6).

7.8   Multiecho Sequences

Several echoes can be generated in a single cycle with both SE and GRE sequences: additional spin echoes are produced by applying extra 180° refocusing RF pulses while multiple gradient echoes are generated by repeat reversal of the frequency-encoding gradient. Multiecho techniques are employed for two reasons:

— The generation of multiple echoes enables acquisition of a sequence with *several measurements that differ in their echo times and T2 weightings*. For instance, a repetition time of 2000 msec with echo times of 20 msec for the first and 80 msec for the second echo allows acquisition of a proton density-weighted image (20 msec) and a T2-weighted image (80 msec) with a single measurement. The multiecho technique is routinely used in the clinical setting (▶ Fig. 32).

— The multiecho technique *accelerates data acquisition* and can be used for ultrafast imaging (▶ Chapter 8).

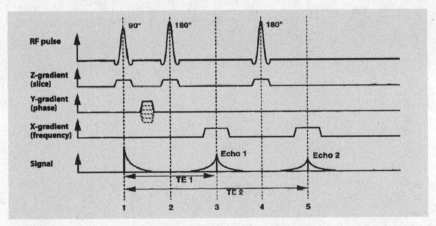

Fig. 32. Multiecho SE sequence. A second 180° refocusing RF pulse (4) is applied to generate a second echo (5), resulting in an image with heavier T2 weighting due to the longer TE. The second 180° pulse is delivered exactly midway between the first (3) and the second (5) echo

References

1. Mitchell DG, Cohen MS (2004) MRI principles, 2nd ed. Saunders, Philadelphia
2. Elster AD (1993) Gradient-echo MR imaging: Techniques and acronyms. Radiology 186:1
3. Haacke EM, Frahm J (1991) A guide to understanding key aspects of fast gradient echo imaging. J Magn Reson Imaging 1:621

8 Fast Pulse Sequences

There are several reasons why it is desirable to speed up scanning.
- A fast sequence allows one to perform dynamic studies, e.g. to track a contrast medium bolus.
- Shorter acquisition is less prone to motion artifacts, which is especially important in uncooperative patients.
- A sequence that is fast enough can be acquired during breath-hold and thus yields images without respiratory artifacts.

Various techniques are available to shorten scan time:
- Use of state-of-the-art gradient and RF systems to full capacity and more effective timing of conventional sequences ([ultra-]fast GRE).
- Sampling of multiple echoes with different phase encodings (FSE, echo planar imaging).
- Incomplete filling of k-space (fractional echo imaging, partial Fourier imaging, rectangular field of view).

8.1 Fast or Turbo Spin Echo Sequences

Fast spin echo (FSE) sequences (also called turbo spin echo (TSE) sequences by some manufacturers) are modified SE sequences with considerably shorter scan times. This is accomplished by delivering several 180° refocusing RF pulses during each TR interval and briefly switching on the phase-encoding gradient between echoes. In this way, optimal use is made of the TR interval by sampling several echoes *with different phase encodings* after each excitation pulse (▶ Fig. 33). The series of spin echoes thus generated is called an *echo train* and the number of echoes sampled is the *echo train length (ETL)*. The imaging time of an FSE sequence is calculated as:

> Scan time = TR × number of phase-encoding steps × number of signal averages [ETL]
>
> ETL is the *echo train length* and refers to the number of echoes sampled per echo train.

FSE sequences are not only faster but differ from conventional SE techniques in a number of other ways as well.

— FSE sequences have a longer TR in order to deliver as many 180° refocusing RF pulses as possible. The TR of FSE is 4000 msec or greater compared with 2000–2500 msec for SE sequences. With their longer TR, FSE sequences are well suited for the acquisition of T2-weighted images.

— The TE of FSE sequences for T2-weighted images is also longer.

The fact that several echoes can be generated after a single excitation pulse is exploited in conventional imaging to acquire a proton density-weighted (intermediate-weighted) image and a T2-weighted image with the same sequence (► Chapter 7.8). Alternatively, the multiecho technique can be used to acquire faster sequences.

FSE sequences can be used to perform double echo imaging by splitting the echo train. With an echo train length of eight, for example, the first four echoes can be used to generate a proton density-weighted image and the last four echoes to generate a T2-weighted image.

8.2   Single-Shot Fast Spin Echo (SSFSE) Sequences

Single-shot fast spin echo (SSFSE) and half-Fourier acquisition single-shot fast spin echo (HASTE) are alternative names for a very fast MR technique with scan times of 1 sec or less. The technique is based on incomplete k-space filling (fractional echo and partial Fourier imaging). "Single-shot" indicates that half of the k-space lines are filled after only one RF excitation pulse. The speed of acquisition reduces motion artifacts to a minimum. Because of the long echo times, SSFSE or HASTE images selectively depict tissues with long TEs, i.e. compartments containing free liquid, whereas tissues with short or medium-length TEs are not shown. For this reason, the SSFSE or HASTE technique is used for MR myelography, MR urography, and MR cholangiopancreatography (MRCP).

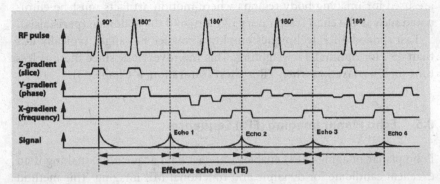

Fig. 33. Fast spin echo sequence. Four 180° refocusing RF pulses are applied to create four echoes (echo train). Since, in contrast to the multiecho technique, the phase-encoding gradient is switched on before each echo, the four echoes obtained after a single excitation pulse have different phase encodings. In the example shown, T2 contrast is determined principally by the third echo (effective TE, ▸ Chapter 8.9)

8.3 Fast or Turbo Inversion Recovery (Fast STIR) Sequences

Modifying the echo trains of an IR sequence is especially effective because the extremely long TRs allow for full T1 relaxation to occur. Fast or turbo inversion recovery (fast STIR) sequences have the same inversion time as conventional STIR sequences and also use an initial 180° inversion pulse but sample all echoes of an echo train with different phase encodings.

8.4 Fast Gradient Echo (GRE) Sequences

Fast gradient echo (GRE) sequences (also known as *turbo gradient echo* or *ultrafast gradient echo sequences*) used in conjunction with state-of-the-art gradient systems (active shielding) achieve echo times below 1 msec with repetition times of 5 msec or less. Fast GRE is basically a conventional GRE sequence that is run faster and uses some mathematical tricks, primarily incomplete filling of k-space (fractional echo and partial Fourier imaging, ▸ Chapter 5.3). Fast GRE sequences yield an excellent image quality although a slice can be acquired in only a few seconds (typically 2–3 sec). Such sequences are highly suitable for dynamic imaging, for example, to track the inflow of a contrast medium bolus. Moreover, fast GRE techniques are used for imaging body regions where motion artifacts must be eliminated such as the chest (respiratory motion) and the abdomen (peristalsis).

Fast spoiled GRE techniques employ a smaller flip angle, typically less than 45°, for optimal T1 weighting. This improves SNR since there is less time for T1 relaxation when TR is short (saturation, ▶ Chapter 3).

8.5　Echo Planar Imaging (EPI) Sequence

Echo planar imaging (EPI) enables ultrafast data acquisition, making it an excellent candidate for dynamic and functional MR imaging. This method requires strong and rapidly switched frequency-encoding gradients. An echo train consisting of up to 128 echoes can be acquired (▶ Fig. 34). In this way, it is possible to obtain an image with a resolution of 256×128 after a single excitation pulse (single shot) in 70 msec, which corresponds to 16 images per second! However, EPI still has to tackle a couple of problems, which have so far precluded its routine clinical use. These are:
— As a GRE technique, EPI cannot compensate for field inhomogeneities and the signal decays with T2*.
— The rapidly switched gradients induce field inhomogeneities that accumulate over time, causing geometrical distortions of the MR image.
— Due to rapid T2* decay of the signal, there is only little time for echo collection. To perform an adequate number of measurements in the short interval available, a very strong and fast gradient is needed. The speed of gradient switching is limited by the electrical inertia of the gradient coils and by the risk of damage to the person being imaged as a result of nerve stimulation associated with rapidly changing magnetic fields. Moreover, rapid gradient switching is so noisy that patients need ear protection!
— Image contrast is often rather poor since a single-shot acquisition involves no repetition and hence there is no T1 effect. Contrast can be improved by applying a presaturation pulse but only at the expense of the signal-to-noise ratio, which is already poor.

8.6　Hybrid Sequences

Hybrid techniques generate and record a series of alternating SEs and GREs. GRASE (gradient and spin echo) and spiral imaging are hybrid techniques.

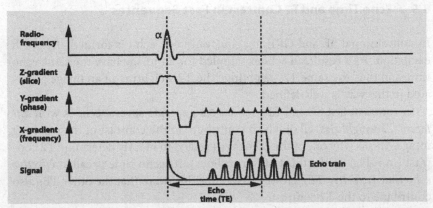

Fig. 34. Echo planar imaging (EPI). As with the FSE technique, several echoes (eight in the example shown) are generated with different phase encodings. In contrast to FSE, the echoes are not generated with a 180° RF pulse but with the frequency-encoding gradient – as in a GRE sequence. This technique requires powerful amplifiers since the frequency-encoding gradient must be reversed very rapidly. The peaks of the phase-encoding gradient are called "blips"

8.7 Gradient and Spin Echo (GRASE) Sequence

A gradient and spin echo (GRASE) sequence is a combination of FSE and EPI. A series of 180° RF pulses is applied to generate several spin echoes (as in FSE). In addition, several GREs are produced for each SE by rapidly switching the readout gradient polarity. This makes the GRASE technique even faster than FSE without impairing image quality as the signal decays with T2 rather than with T2*. The contrast achieved is the same as that obtained with conventional SE sequences.

8.8 Spiral Sequences

Spiral sequences derive their name from the fact that k-space is filled using a spiral trajectory. Spiral imaging is performed with a GRE sequence combined with two oscillating gradients. It is a promising approach, especially for real-time imaging of the heart.

8.9 Echo Time and T2 Contrast in Fast Sequences

In conventional SE and GRE imaging, only one echo is formed after each excitation. As a result, all echoes sampled for an image have the same echo time and thus the same T2 weighting. The T2 weighting of an image generated in this way is well defined.

In contrast, fast SE and EPI sequences generate several echoes with *different T2 weightings*, all of which contribute to the contrast of the resulting image. This is why one of the echoes is selected to mainly determine T2 contrast (in ► Fig. 33 the third of four echoes). Its echo time is called *effective echo time (effective TE)*. However, we must be aware that the other TEs also contribute to the T2 contrast.

Technically, the echo is selected by recording it in such a way that it fills the center of k-space (► Chapter 4.2), which contains the data that most strongly affect image contrast.

References

1. Elster AD (1993) Gradient-echo MR imaging: Techniques and acronyms. Radiology 186:1
2. Frahm J, Häenicke W (1999) Rapid scan techniques. In: Stark DD, Bradley WG Jr (eds) Magnetic resonance imaging, 3rd edn. Mosby-Year Book no 87, Mosby, St. Louis

9 Fat Suppression Techniques

Several techniques are employed in clinical MR imaging to reduce (suppress) the signal from fat.
— Chemical shift imaging based on the time-dependent phase shifts between water and fat
— Frequency-selective fat saturation (fat sat pulse)
— T1-dependent fat suppression (STIR)
— Spectral presaturation with inversion recovery (SPIR)

9.1 Chemical Shift Imaging

As already mentioned, the same atomic nucleus differs slightly in its resonant frequency when bound in different molecules or at different molecular sites. This type of resonant frequency difference is known as *chemical shift*. The chemical shift can be given in Hertz (Hz), which is proportional to the strength of an external magnetic field to which the protons are exposed, or as "parts per million" (ppm), a unit which is independent of the magnetic field strength.

The chemical shift most important in clinical imaging is that between protons in fat and water. The resonant frequency of fat protons bound in long-chained fatty acids (e.g. triglycerides) and water protons differs by 3.5 ppm, which, at a field strength of 1.5 T, causes fat to precess 225 Hz slower than water (▶ Fig. 35). If the water and fat protons are in the same voxel, the precessional frequency difference will become apparent as a phase difference after magnetization has been tilted into the xy-plane and transverse relaxation has occurred. Over time, fat and water protons fall alternately *in* and *out of phase* with each other. They are said to be in *opposed phase* when their phase difference is 180°. At 1.5 T fat and water protons will be

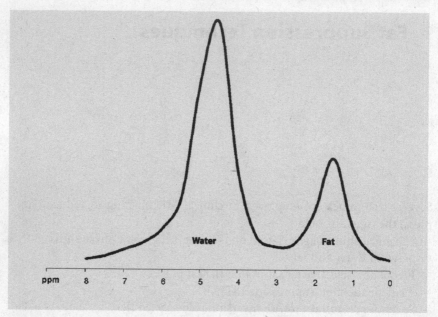

Fig. 35. Chemical shift between fat and water. The resonant frequencies of fat and water protons are separated by approximately 3.5 ppm, which translates into a difference of 225 Hz at 1.5 T

180° out of phase 2.2 msec after excitation and in phase again after 4.4 msec. After another 2.2 msec they will again be out of phase and so on. In clinical MR imaging, these time-dependent phase shifts between the two protons are exploited to suppress the fat (or water) signal selectively. In an image acquired under in-phase conditions, the transverse magnetization components of water and fat protons which are in the same voxel add together and produce a strong signal, while in an out-of-phase image either water or fat alone contributes to the signal (▶ Fig. 36). The differences in signal intensities between in-phase and opposed-phase images can help differentiate benign and malignant lesions in clinical MR imaging. If an organ lesion contains fat, this will cause a decrease in intralesional signal intensity on the opposed-phase image compared with the in-phase image. This technique is known as *chemical shift imaging* and, for example, has a role in the MR evaluation of adrenal tumors, where the presence of fat is an important criterion for lesion characterization.

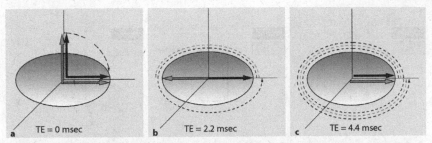

Fig. 36a–c. Phase differences between fat (gray arrow) and water (black arrow) as a function of echo time (TE). At an external magnetic field strength of 1.5 T, the transverse magnetization vectors of fat and water point in opposite directions at TE = 2.2 msec (**b**), resulting in a weak MR signal. (**c**) At TE = 4.4 msec, water and fat are back in phase and both contribute to the MR signal

A technique of chemical shift imaging for the selective suppression of the signals from either fat or water was proposed by *Dixon*. In this method two sets of images are acquired, one with fat and water signals in phase and the other with fat and water signals out of phase. The signal intensities of the two images obtained with this method (image 1 and image 2) can be described as:
— Image 1 = water plus fat
— Image 2 = water minus fat
By adding image 1 and image 2, a pure water image (water plus water) is reconstructed while subtracting image 2 from image 1 generates a pure fat image.

9.2   Frequency-Selective Fat Saturation

Because water and fat have different resonance frequencies, it is possible to selectively saturate the spectral peak of either water or fat by applying a frequency-selective RF pulse before imaging. "True" saturation methods deliver the RF pulse after calibration has been performed to exactly determine the spectral peak of fat. These methods are frequently used in MR spectroscopy but not in routine clinical MR imaging, where fat suppression is generally accomplished by means of a *spoiling* technique. A fat sat pulse is a short frequency-selective 90° RF pulse that is applied to rotate the fat magnetization into the transverse plane. While in the transverse plane, the

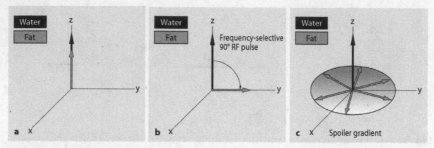

Fig. 37. Frequency-selective fat suppression. A frequency-selective 90° RF pulse is applied to rotate the fat magnetization vector into the transverse plane (**a, b**). The fat spins begin to dephase, which is accelerated by applying a spoiler gradient. Thus, only the longitudinal magnetization of water is available for subsequent excitation (**c**)

fat magnetization is dephased by application of a spoiler gradient, leaving only the longitudinal magnetization of water for excitation during the next cycle (▶ Figs. 36 and 37).

Frequency-selective fat suppression techniques are typically used on high-field scanners while STIR sequences are preferred on low-field scanners.

9.3 Short TI Inversion Recovery (STIR)

STIR sequences provide reliable fat suppression at all field strengths. They are mainly used for fat suppression on low-field scanners and in all other instances where adequate fat suppression cannot be achieved by means of frequency-selective techniques. The principal function of the STIR sequence is described in ▶ Chapter 7.5.

9.4 Spectral Presaturation with Inversion Recovery (SPIR)

SPIR is similar to STIR in that it is an inversion technique for fat suppression. However, while the STIR sequence uses an initial 180° saturation pulse, the SPIR technique employs an initial inverting pulse that is made frequency-selective and only inverts fat magnetization. Note that SPIR is not a pulse sequence but merely an additional module that can be applied prior to other pulse sequences. The SPIR module is typically used to obtain fat-suppressed images in conjunction with a T1-weighted sequence.

10 Parallel Imaging

KLAAS P. PRUESSMANN

10.1 Background

The fast MR sequences presented in the preceding chapters are basically conventional sequences that are run faster. With these sequences, much shorter scan times are achieved but the extent to which data acquisition can be accelerated in this way is limited by the available hardware, in particular the performance and slew rates of the gradient coils used for frequency and phase encoding. Moreover, the use of ever more powerful gradients with higher slew rates is limited by physiological considerations such as the risk of peripheral nerve stimulation.

Another concern is that RF energy deposition in a tissue leads to heating (specific absorption rate, SAR). To ensure patient safety, SAR limits have been defined for MR imaging. These limits may be exceeded when fast imaging protocols with large flip angles or extremely short repetition times are used.

10.2 Principles of Parallel Imaging

Parallel imaging methods offer an interesting solution to the limitations just outlined. These techniques use a set of surface coils placed side by side for the simultaneous acquisition of several reduced data sets. Such a multiple receiver coil array allows for further shortening of acquisition time but in a way that is fundamentally different from the techniques used in conventional fast sequences. In parallel imaging, scan time is shortened by reducing the number of phase-encoding steps rather than by further speeding up the succession of steps. The desired scan time reduction is thus achieved without faster gradient switching rates and without increasing the risk of tissue heating.

Specifically, the number of phase-encoding steps is reduced by incomplete sampling of k-space. When k-space is filled less densely by collecting fewer phase-encoding spin echoes, a linear reduction in scan times results because image acquisition time is proportional to the number of phase-encoded echoes collected. For example, scan time can be reduced by 50 percent if only every other line is filled. The immediate effect, however, is an undesired one, namely a smaller field of view in the phase-encoding direction (▶ Fig. 38) and the occurrence of wraparound artifacts. This means that parts of the imaged volume that extend beyond the FOV are spatially mismapped to the opposite side of the image.

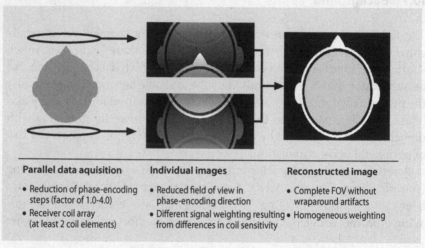

Parallel data aquisition	Individual images	Reconstructed image
• Reduction of phase-encoding steps (factor of 1.0–4.0)	• Reduced field of view in phase-encoding direction	• Complete FOV without wraparound artifacts
• Receiver coil array (at least 2 coil elements)	• Different signal weighting resulting from differences in coil sensitivity	• Homogeneous weighting

Fig. 38. In parallel imaging, an array of receiver coils simultaneously collect the MR signals. Scan time is shortened by reducing the number of phase-encoding steps. As a result, the individual images are obtained with a smaller FOV and show the typical wraparound artifacts. A complete image without wraparound artifacts is reconstructed by combining the individual images

These wraparound artifacts can be eliminated by using parallel imaging. In this technique, each element of the array of coils yields a separate image with a small FOV where part of the image information is obscured by wraparound artifacts. However, the superimposed portions are characterized by different weightings that vary with the spatial sensitivity of the respective coil element. In ▶ Fig. 38, for example, the coil placed in front is more sensitive to the face and the coil placed behind primarily only images the back of the head. Knowledge of these individual sensitivities allows mathematical separation of the information underneath and reconstruction of an image comprising the overall FOV without wraparound artifacts. Moreover, the reconstruction process also eliminates the different weightings, resulting in a final image of homogeneous signal intensity.

10.3 Special Requirements

With respect to hardware, the most important item necessary to perform parallel imaging is a suitable array of receiver coils. Depending on the intended application, the array of coils consists of two to eight elements. Proper geometric arrangement of the coils is crucial for the signal-to-noise ratio (SNR) attainable. It is also important to keep the spatial sensitivities of the array elements fairly constant during imaging. This is accomplished by a rigid arrangement, for instance, a cage-like configuration when the head is imaged. In contrast, flexible arrangements that can be attached to the patient in an individual manner are preferred for parallel imaging of the chest and abdomen. Finally, the MR scanner must have a corresponding number of separate receiver channels to connect each of the coil elements.

> To ensure reliable image reconstruction in parallel imaging, it is important to precisely determine the coding effects of the individual receiver sensitivities. This is often done by performing an additional reference measurement at the beginning of each examination (calibration). Alternatively, individual reference data can be acquired with each image acquisition.

10.4 Applications

Parallel imaging can be used to shorten acquisition time in conjunction with virtually all known sequences and contrast mechanisms. As a rule, parallel acquisition does not alter the contrast characteristics and therefore the images can be interpreted in the same way as their conventional counterparts.

The gain in speed is directly proportional to the reduction of phase-encoding steps. The *acceleration factor* is the factor by which the number of sampled k-space lines is reduced. It can take on any whole-number or fractional value between 1.0 (no acceleration) and about 3.0 to 4.0. Even faster data acquisition is possible with 3D techniques that achieve further acceleration by virtue of their two phase-encoding directions.

Commercially available parallel imaging software is marketed as SENSE, IPAT, ASSET, or SPEEDER. The faster scan time achieved with these tools is of use in a wide range of practical applications. In the clinical setting, a reduction of scan time is especially attractive for imaging protocols with very long sequences or imaging during breath-hold. Short scan times are also beneficial in dynamic MR studies such as evaluation of contrast medium passage or cardiac motion. Alternatively, parallel imaging techniques can be employed to improve spatial resolution or to acquire more slices without unduly increasing scan time.

Finally, parallel imaging can help reduce artifacts. When sequences with long acquisition times are used, shorter readout trains can reduce undesired effects that interfere with image quality. This applies especially to echo planar imaging (EPI), which is frequently degraded by considerable artifacts caused by field inhomogeneities due to variable susceptibility, movement, and flow. Moreover, the extremely rapid gradient reversal necessary in EPI is associated with a very high noise level. Parallel imaging is less noisy because the gradient reversal rate is reduced by shortening the readout train while the overall scan time remains the same.

Whenever one considers applying a parallel imaging technique for any of the reasons outlined, one should also be aware that the sequence used should have some SNR reserve. This is necessary because, with few exceptions, parallel imaging will reduce SNR.

References

1. Sodickson DK, Manning WJ (1997) Simultaneous acquisition of spatial harmonics (SMASH): fast imaging with radiofrequency coil arrays. Magn Reson Med 38:591–603
2. Pruessmann KP, Weiger M, Scheidegger MB, Boesiger P (1999) SENSE: sensitivity encoding for fast MRI. Magn Reson Med 42:952–962
3. Griswold MA, Jakob PM, Heidemann RM, Nittka M, Jellus V, Wang J et al (2002) Generalized autocalibrating partially parallel acquisitions (GRAPPA). Magn Reson Med 47:1202–1210

11 Cardiovascular Imaging

DANIEL NANZ

The cardiovascular system can be examined by MR imaging at different levels.

Vessels are depicted directly (MR angiography, MRA) and can be evaluated for anatomic abnormalities, narrowing, dilatation, or dissection. The advent of relaxation contrast media has dramatically changed vascular MR imaging and has in particular facilitated time-resolved studies. MR images can depict not only the blood but also the vessel wall and its diseases.

Blood vessels and capillaries with diameters well below 1 mm are usually not seen directly. However, with dedicated techniques, the MR signal from tissues can be made to vary in proportion to the blood flow in their capillary beds. In this way it is possible to directly visualize relative regional differences in organ perfusion.

Effects of *perfusion deficits* caused by a stroke can be evaluated on diffusion-weighted MR images obtained within minutes of the onset of symptoms. On such images, the signal intensity reflects the mobility of water molecules at the microscopic level.

In the brain, functional MR imaging provides indirect information on cerebral activity by depicting changes in the *oxygen saturation* of the capillary blood.

MR imaging of the heart presents some specific problems. Notwithstanding, a wide range of clinical questions can be answered by a set of MR images of the *myocardium* or heart muscle obtained with a combination of different sequences.

References

1. Clinical indications for cardiovascular magnetic resonance (CMR) (2004) Consensus Panel report. European Heart Journal 25:1940–1965

11.1 Angiography

Angiographic MR imaging techniques have been optimized to image the blood and surrounding anatomy with different signal intensities. For three of the techniques presented below (time-of-flight, phase-contrast, and black blood MRA), successful differentiation requires that blood move faster than surrounding structures. The fourth technique, contrast-enhanced MRA, is different in that it images body regions with a bright signal if their local longitudinal relaxation times are shortened to values below 100 msec by the presence of an externally administered contrast agent. After direct injection of the agent into the vascular system, this condition is initially only met within the vascular tree, which allows selective enhancement of the blood signal.

11.1.1 Bright Blood Imaging

The MRA techniques most widely used in the routine clinical setting depict the blood with a high signal intensity (bright blood imaging). Vessels with positive contrast are more conspicuous and, in electronic postprocessing of MRI data, can be more easily visualized on projections through stacks of images. However, all bright blood techniques are limited by the fact that there is usually no signal from blood when flow is turbulent. Under these conditions, the blood cannot be distinguished from surrounding tissue. Turbulent flow often occurs in important vessel segments such as branchings or vessel segments distal to a stenosis. In general, the only remedy to reduce this effect is to keep the echo time as short as possible.

Angiographic MR techniques can be used to acquire two-dimensional (2D) or three-dimensional (3D) data sets. 2D data can be postprocessed to generate 3D volumes. A general advantage of 3D imaging is that thinner slices can be obtained without interslice gaps. Moreover, volumetric data sets allow for multiplanar reformation with good resolution. When MRA is performed in the 2D mode, optimal results are achieved with the slices oriented perpendicular to the vessel of interest and scanning against the direction of blood flow. This will minimize undesired saturation and partial volume effects (▶ Fig. 39).

Time-of-Flight (TOF) MR Angiography

Time-of-flight (TOF) MR angiography images blood with a high signal intensity (bright) if it rapidly flows through the imaging plane. TOF angiog-

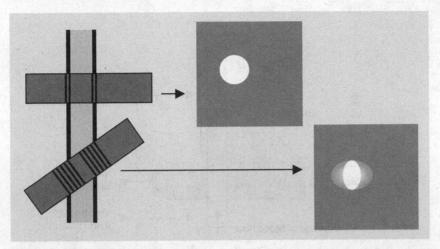

Fig. 39. Partial volume effects are accentuated when the imaged slice is tilted out of the plane perpendicular to the longitudinal vessel axis. The vessel diameter appears smaller on the image from the tilted plane

raphy is mainly performed in axial orientation for evaluation of the vessels of the head and neck such as the carotid arteries and the circle of Willis. However, TOF MR angiography is still a valuable option for vascular imaging throughout the body.

The term "time of flight" is probably adopted from a mass spectrometry technique that separates molecular fragments with different masses on the basis of the different times needed by the fragments to travel through a vacuum tube. In a similar manner, TOF angiography depicts the spins of water molecules that move in the blood through the vessels. A vessel appears bright when there is a continuous supply of "fresh" spins that replace the spins in the imaging plane (inflow effect, ▶ Fig. 40).

TOF MRA is performed using GRE sequences with short repetition times (30–50 msec). Echo times should be kept as short as possible. The flip angles used range from approximately 20–40° for 3D imaging to 50° or even greater for 2D imaging. The spins that rest within the slab without moving are highly saturated by the repeated excitation pulses (▶ Figs. 11 and 12) and give only a very weak signal, making stationary tissue appear dark on the resultant image. In contrast, blood flowing into the imaging plane has not been subjected to these RF pulses. As a result, a larger longitudinal magnetization is available for subsequent excitation and the inflowing blood appears bright.

If the newly arriving spins do not leave the scan volume within one TR interval, their magnetization will also be saturated by subsequent RF ex-

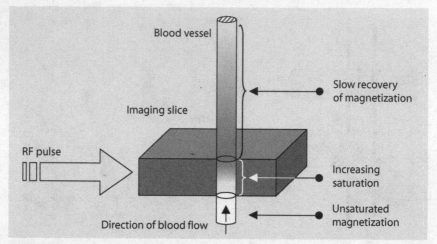

Fig. 40. Principle of TOF angiography. Different gray scale values represent the magnitude of longitudinal magnetization

citation pulses. The MR signal thus becomes smaller and smaller as the spins move away from the entry point into the imaging volume. The longer blood stays in the imaged volume, the smaller is the signal difference versus background. This may give rise to problems in case of slow flow (caused by vascular pathologies, such as aneurysms, false lumina, or vascular malformations, or by the reduction of flow velocity near the vessel wall), in case of a vessel taking a curved course through the slice, or in case of thick-slab acquisitions (especially in 3D imaging). The increasing signal loss can be mitigated to some extent by gradually increasing the flip angle that is experienced by the spins on their way through the scan volume (tilted optimized non-saturating excitation, TONE). Alternatively, a thick slab can be subdivided (multiple overlapping thin-slab acquisition, MOTSA).

Maximal enhancement of flow occurs when thin 2D slices are acquired perpendicular to the direction of flowing blood. Thus, 2D MRA techniques may offer advantages in imaging vessels with slow flow such as the portal venous system.

Problems with magnetization saturation may also be encountered when a vessel does not take a straight course but leaves the scan plane and then enters it again. This may lead to a very weak blood signal in distal vessel segments.

The increase in signal induced by inflowing blood is independent of the direction from which the blood enters the imaging plane. For this reason veins are not readily distinguished from arteries in TOF MRA. This problem can be overcome by applying regional presaturation pulses prior to data acquisition with the goal of achieving complete saturation of magnetization either in a slice distal to the imaging slice (arteriography) or proximal to it (venography). Blood flowing from the presaturated slice into the scan volume appears dark (▶ Fig. 41).

The signal from stationary tissue can be suppressed further by saturating the magnetization of the pool of bound protons (▶ Chapter 3.6), which will improve vessel contrast in many instances. Fat suppression is another option which may improve contrast in special cases.

The presence of moderate concentrations of an MRA contrast medium increases the vessel signal but differentiation of arteries and veins will be more difficult.

The radiologist interpreting TOF MRA images must be aware that the vessel diameter is typically underestimated while a stenosis tends to be overestimated and that contrast may be poor when there is slow blood flow or the vessels do not take a straight course. Also, an unexpectedly bright signal may be seen when relaxation times are shorter than usual. This might be the case, e.g., in the presence of unusually high concentrations of methemoglobin, as they may occur in a hematoma or thrombus.

Advantages of TOF angiography are its robustness under routine clinical conditions and efficient data acquisition.

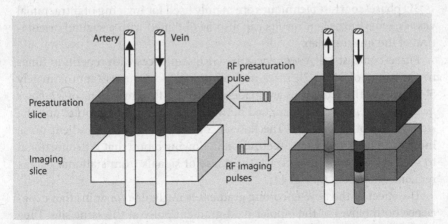

Fig. 41. Differentiation of arteries and veins in TOF angiography. After presaturation of the blood on either side of the imaging slice, the signal intensity of a vessel depends more strongly on the direction of blood flow

Phase-Contrast Angiography

Phase-contrast (PC) angiography images blood with high signal intensity if it flows in the direction of a magnetic field gradient, which is temporarily generated by the MR scanner. The operator can control both the flow direction and the range of flow velocities to which the sequence is sensitive by selecting the direction and the amplitude of the field gradient, respectively. In PC angiography, an average flow velocity can be quantitatively determined for each voxel in the imaging volume.

2D images acquired upstream and downstream of a stenosis, e.g. in a renal artery, can be used to estimate the pressure drop over the stenotic vessel segment. A slice through the stenosis allows one to determine peak flow velocity and the degree of luminal narrowing.

In cardiac imaging, a 2D slice positioned in the ascending aorta just above the aortic valve will provide information on the distribution of outflow velocities over the cross-sectional area of the aorta for "all" (e.g. 20) ECG phases of a cardiac cycle. To this end, a series of 2D phase-contrast angiograms is acquired in synchronization with the heart rhythm at different times during the cardiac cycle (cine phase-contrast imaging). From such a data set, the stroke volume and cardiac output can be estimated. Moreover, an incompetent aortic valve can be diagnosed and the insufficiency quantified by determining the regurgitation volume relative to the stroke volume. Such a velocity profile can also provide information on the shear forces acting on the vessel wall.

3D phase-contrast techniques are mainly used for imaging of intracranial vessels, where excellent results can also be obtained with a sagittal orientation of the imaging slab.

Phase-contrast MRA sequences are GRE sequences with repetition times in the range of 10 to 20 msec and minimum echo times (approximately 5–10 msec). The sequences are made sensitive to flow by means of a *bipolar field-gradient pulse* that is applied between the RF excitation pulse and the signal readout (► Fig. 42). The flow-encoding bipolar field-gradient pulse induces a phase shift of magnetization in flowing blood that is proportional to the flow velocity. In contrast, the phase of signals from stationary spins remains unaffected (► Fig. 43).

The effect of the flow-encoding gradient is negligible for spins that experience both halves of the bipolar field-gradient pulse at the same site. They are subject to a change of their Larmor frequency during the first half of the bipolar pulse, as a result of the change in local magnetic field strength, and they thus do precess at a different rate. However, the second half of the

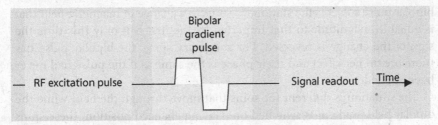

Fig. 42. Diagram of a PC MRA sequence

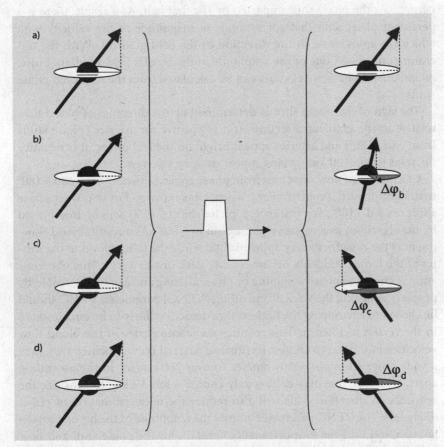

Fig. 43a–d. Sketch of phase shifts induced by a bipolar field-gradient pulse for magnetization associated with stationary spins (**a**, $\Delta\phi_a = 0$), spins slowly flowing in the direction of the gradient field (**b**, $\Delta\phi_b > 0$), spins rapidly flowing in the direction of the gradient field (**c**, $\Delta\phi_c > \Delta\phi_b$), and spins rapidly flowing in the opposite direction (**d**, $\Delta\phi_d = -\Delta\phi_c$). In a phase-contrast image, the gray scale value of a pixel represents the averaged difference angle or phase shift, $\Delta\phi$, measured in the corresponding voxel

bipolar pulse subjects the stationary spins to a change in magnetic field that is equal in magnitude to that imparted by the first half only this time the sign of the change is reversed. For stationary spins, the bipolar pulse has therefore no net effect and their phase is the same as if the pulse had never been applied.

The situation is different for spins that move through the field while the bipolar field gradient is switched on. Having changed position, these spins are exposed to a field change of a different magnitude as well as a reversed sign during the second half of the pulse. This field change cannot fully compensate for the phase shift imparted by the first half. As a result, there is a persistent phase shift that corresponds in magnitude to the velocity with which the spins move in the direction of the field gradient. With the operator-controlled value of the amplitude of the bipolar field-gradient pulse, *quantitative blood flow velocities* can be calculated from the observed phase shifts.

The sign of the phase shift is determined by the direction of blood flow relative to the gradient direction. If it is positive for arteries (phase shifts from 0 to +180°) and arteries appear bright on the MR image, it is negative for veins (0 to −180°) and veins appear dark, or vice versa.

Calculation of flow velocities from phase angles between −180° and +180° is straightforward. Problems arise when spins move so fast that their phase shifts exceed +180°. For instance, a phase shift of +200° will be interpreted by the algorithm as a negative phase shift of −160°. As a result, blood flowing near the vessel wall may appear bright while the faster blood in the center of the lumen suddenly becomes quite dark or vice versa. This phenomenon is known as phase wrapping or phase aliasing and can be prevented by properly adjusting the velocity encoding (VENC) parameter. VENC should be chosen to encompass the highest flow velocities likely to be encountered in the vessels of interest. This requires some knowledge of the blood flow velocities in different vascular territories. Arterial flow velocities vary over a wide range from just a few cm/sec to over 200 cm/sec in the ascending aorta. However, one may deliberately choose a low VENC to sensitize the sequence to slow flow. This will also reduce the underestimation of vessel diameters. The VENC parameter adjusts the amplitude of the bipolar gradient pair and thus the proportionality constant for the phase shift and flow velocity.

The absolute phase of an MR signal is affected by numerous factors. Some of these are ill defined but well reproducible on a time scale of seconds. To isolate phase changes caused by the bipolar field-gradient pulse, phase-contrast MRA methods collect two data sets, which only differ in the shape of the bipolar pulse. For the acquisition of the second data set either the amplitudes of the two halves of the pulse are reversed (a +/– pulse followed by a –/+ pulse or vice versa) or both are set to zero. A systematic error in the phase measurement can thus be corrected by subtraction of the two data sets. The tradeoff, however, is a longer minimum scan time.

After the acquisition of four data sets with systematic variation of the amplitudes of the two halves of the flow-sensitizing bipolar field-gradient pulse, the three components of the flow velocity vector can be measured with error correction. MR angiograms generated in this way resemble those obtained with other angiographic techniques, since they depict flowing blood bright. However, in contrast to angiograms acquired with other techniques, the blood signal intensity in these phase-contrast angiograms only depends on the flow velocity and is not affected by the flow direction.

Moderate amounts of an MRA contrast medium will increase the signal intensity of blood and thus improve SNR.

The acquisition of 3D phase-contrast angiograms with flow encoding in all three spatial directions can be time consuming. Fast flowing blood in large arteries and nearly stagnant blood in an aneurysm or vascular malformation cannot both be adequately depicted with a high sensitivity in a single measurement. Like other techniques, phase-contrast angiography also tends to underestimate vessel diameters and overestimate stenoses.

Advantages of the phase-contrast technique include the quantitative and spatially resolved evaluation of flow velocities and flow directions and the good suppression of the signal from stationary tissue. With proper parameter settings, phase-contrast MR angiography is most suitable for depicting slow flow or flow within the imaging slice. No other MR technique provides the kind of quantitative information that can be derived from the time-resolved velocity and flow profiles that can be obtained with cine phase-contrast angiography for different phases of the cardiac cycle.

Contrast-Enhanced MR Angiography

Contrast-enhanced MR angiography images blood with a bright signal if its longitudinal relaxation time is sufficiently shortened by the presence of a relaxation-enhancing contrast medium (► Fig. 44). Contrast-enhanced MR angiography enables rapid acquisition (within seconds) of three-dimensional data sets with a good SNR and a resolution in the millimeter range

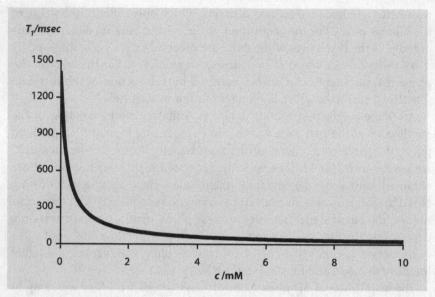

Fig. 44. Shortening of the T1 of water in blood by increasing contrast medium concentrations. Approximation for a contrast medium with a molar relaxivity of 4 l/(mmol · sec) and a T1 in the absence of the contrast medium of 1.4 sec

thereby allowing imaging of large segments of the vascular system in all regions of the body. Contrast-enhanced MRA is well established for imaging of the large vessels of the trunk and in the periphery but is also used, in combination with other techniques, for evaluating the vessels of the head and neck.

In general, the contrast agents for MR angiography are injected into a vein in the bend of the elbow. The MRA contrast agents are well-tolerated, gadolinium-based paramagnetic compounds (▶ Chapter 12) that are administered at doses of 0.05–0.3 millimol gadolinium per kilogram of body weight. In arteriography, where a large arterial signal and a small signal from veins are desirable, images must be acquired during the first pass of the contrast medium through the arteries. Conspicuity of the arterial system decreases quickly due to the subsequent signal increase in the veins and perfused tissue. Except for the brain, there is a rapid diffusion of most contrast media through the capillary walls into the extravascular extracellular space. An imaging window of only a few seconds is available between the arterial inflow of the contrast medium and its arrival in the veins. Therefore, the timing of the acquisition start and the duration of the scan are of prime importance in contrast-enhanced MR angiography. Typical scan times are in the range of 20 seconds, which allows imaging of the thoracic and abdominal vessels during breath-hold.

The scan times can be shortened even further to repeatedly image a body region with update rates that are high enough to track the progressive three-dimensional distribution of the contrast medium in a time-resolved manner.

Another option is to move the scan volume in synchronization with the advancing contrast medium bolus so as to cover a larger body region with several acquisitions (so-called *multi-station bolus chase*). The position of the table carrying the patient is automatically adjusted during scanning. Electronic postprocessing can be used to generate an overview image of the entire anatomy that was sequentially imaged. Under ideal conditions, this technique allows scanning of the arterial system from head to ankle following a single, optimized contrast medium injection. The well-tolerated MR contrast media available today, however, can also be injected repeatedly in a single examination, which provides more flexibility for protocol optimization.

Contrast-enhanced MRA is performed using spoiled GRE sequences with very short TRs (approximately 1.7–6 msec) and very short TEs (below 2 msec). Flip angles ranging from 15° to 50° are used. The sequences closely resemble those used for TOF MRA but with a further marked reduction of repetition and echo times. As a result, the signal from stationary spins in the scan volume is suppressed even more efficiently. In contrast, blood magnetization quickly recovers in the presence of an adequate concentration of a contrast medium (on the order of about 5 millimol/liter, depending on the agent administered and the injection protocol). Blood will then generate a strong signal and appear bright despite the RF excitation pulses that are applied in rapid succession. Occasionally, image contrast can be further enhanced by combining the technique with optimized fat saturation techniques.

Scan time is a crucial issue and all kinds of tricks are applied to shorten image acquisition – most often at the expense of an SNR reduction. Examples include:

— Shortening of echo and repetition times through incomplete readout of the data in the frequency-encoding direction (fractional echo imaging, ► Chapter 5.3).

— Reduction of the number of phase-encoding steps through incomplete readout of the phase-encoding data (partial Fourier imaging, ► Chapter 5.3). The missing data is estimated on the basis of symmetry relations in k-space.

— Reduction of the minimum echo and repetition times by a broader receiver bandwidth.
— Parallel imaging (▶ Chapter 10) with use of suitable receive coil arrays allows a further reduction of the number of phase-encoding steps or the acquisition of images with an improved spatial resolution in the same scan time.

Finally, the effective "exposure time" of the acquisition when imaging large arteries can be even further reduced by an optimized temporal ordering of the phase-encoding steps. The signal intensity and contrast of the bright vessels are largely determined by the data in the center of k-space (Chapter 5.3) (▶ Fig. 45). Phase-encode ordering can be designed to acquire all of the central k-space data at the beginning of the scan by initially applying only phase- and slice-encoding field-gradient pulses of low amplitudes. Such accelerated sampling of the center of k-space while all of the contrast medium is still in the arterial system can produce images of large vessels with good arterial contrast and minimal venous overlap. This even applies when the peripheral k-space data are acquired after a considerable amount of the contrast medium has reached the veins. With centric k-space sampling, longer total scan times and improved image quality are possible without compromising arterial contrast. Commercially available implemen-

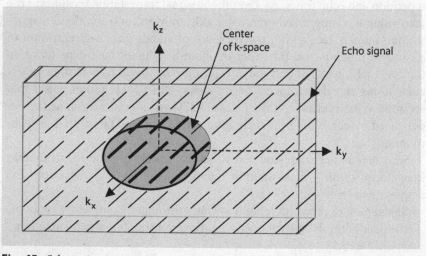

Fig. 45. Schematic representation of the MR raw data in k-space for a three-dimensional image. Each diagonal line represents an echo signal that is recorded within 1 or 2 msec. The contrast of the resulting MR image is mainly determined by the data in the center of k-space

tations of this technique are known as CENTRA or elliptical centric phase-encode ordering.

It is crucial to collect the central k-space lines when the contrast medium concentration in the target vessels is highest. Several strategies are available for *synchronizing data acquisition with bolus passage*:

— The test bolus technique is a method in which the individual patient's circulation time is determined by measuring the time the contrast medium needs to pass from the site of injection to the target vessel. To this end, a small amount of contrast (1 to 2 ml followed by a saline flush) is injected and the target area is repeatedly imaged using a fast sequence, e.g. a spoiled T1-weighted 2D GRE sequence which updates images once every second. The bolus must be large enough to cause signal enhancement when it arrives in the target vessel but should not unduly enhance the background signal in the subsequent 3D data acquisition. Based on the knowledge of the individual circulation time determined in this way and the method of phase-encode ordering used, the start of the 3D angiography sequence can be optimally coordinated with the injection of the contrast medium.

— Automatic triggering techniques are based on the continuous measurement of the vascular signal in a proximal test volume. Tracking starts with the injection of the angiographic bolus of the contrast medium and the 3D sequence is then automatically triggered with an operator-controlled delay as soon as the signal intensity in the region of interest increases above a defined threshold. When the renal arteries are imaged, the test volume can be placed in the abdominal aorta.

— In a similar manner, 3D acquisition can be started manually as soon as the operator observes the arrival of contrast in the target volume on rapidly updated 2D images. This method is occasionally referred to as *fluoroscopic triggering.*

Automatic or manual triggering techniques provide images with optimal arterial contrast when combined with a phase-encode ordering technique that samples the central lines of k-space first. These techniques can be quite sensitive to early or late mistriggering of data acquisition. Moreover, rapid instruction of the patient is needed if breath-hold imaging is necessary. Bolus timing, on the other hand, is readily compatible with any method of k-space filling.

The background signal can often be even better suppressed when the angiographic data set is acquired twice with the same parameters before and after contrast medium injection. The unenhanced image, the so-called *mask*, is then subtracted from the contrast-enhanced image. The resultant difference images highlight the signal changes occurring after contrast medium administration, as long as the patient did not move between the two scans.

Many studies have shown contrast-enhanced MRA to have a high diagnostic accuracy in comparison with conventional radiographic techniques or other reference modalities. Most of the problems that arise in routine clinical application are associated with proper timing of data acquisition relative to contrast medium injection. The problems may be merely technical in nature or due to individual variations in circulation times and contrast medium distribution. An aneurysm, false lumen, or arteriovenous malformation may not be completely filled with contrast medium at the time of scanning even when there is adequate enhancement of the rest of the arterial system. When multi-station bolus chase is used, confounding signals from bright veins on projections of the lower leg images may limit the evaluation of the arterial tree. This is a problem more likely to occur in patients with diabetes mellitus. Retrograde inflow of the contrast medium may impair the diagnosis of vascular occlusion. However, as with other techniques, contrast-enhanced MRA generally overestimates rather than underestimates stenoses. With time-resolved imaging (▶ Chapter 11.1.3), many of the problems associated with achieving optimal bolus timing can be overcome. Contrast medium injection is minimally invasive. MR contrast media are associated with a very low rate of adverse events and allergic reactions are rare (▶ Chapter 12).

The advantages of contrast-enhanced MR angiography include:
— short scan time,
— three-dimensional display of large volumes in any orientation,
— high SNR and good vessel contrast,
— no exposure to ionizing radiation,
— well tolerated contrast medium,
— minimal invasiveness of contrast medium injection, and
— reasonable robustness of the method under routine clinical conditions.

11.1.2 Black Blood Imaging

Black blood MR angiography images blood vessel lumina with a low signal intensity, if the blood they contain is completely replaced by fresh blood that enters the imaging volume during data acquisition. In contrast to the TOF MRA technique, the inflowing blood does not produce a bright signal but is imaged dark with a low signal intensity.

Black blood MRA sequences are well suited to evaluate the vessel walls and the myocardium. They provide information on wall thickness, the presence of inflammatory wall lesions, and the internal makeup of mural thrombi. So far, black blood MRA techniques have been mainly used to image large vessels such as the thoracic and abdominal aorta and the heart chambers or easily accessible vessels such as the carotid arteries. However, there are also a few examples of black blood coronary artery images.

Black blood angiography is performed using SE sequences whereas TOF imaging is performed with GRE sequences. This explains the different effects of fresh blood flowing into the scan plane, which appears dark in black blood angiography but bright in TOF angiography. Blood whose magnetization is rotated into the transverse plane by the 90° excitation pulse of an SE sequence and then leaves the slice before the 180° refocusing pulse is delivered does not generate a signal (▶ Chapter 7.2). The two pulses are separated by half the echo time. Likewise, there is no signal from blood which is still outside the slice when the 90° RF pulse is applied but which then flows into the slice between excitation and readout.

The blood signal can be suppressed even more effectively by double inversion of longitudinal magnetization some hundreds of milliseconds before data sampling (double inversion recovery, ▶ Fig. 46). In this method, a non-selective 180° pulse, followed by a slice-selective 180° pulse, is applied to selectively rotate only the magnetization outside the scan plane into the negative z-direction. The magnetization relaxes and passes through zero before it regrows in the positive z-direction. Three conditions must be met for an improved suppression of the blood signal by double inversion recovery:

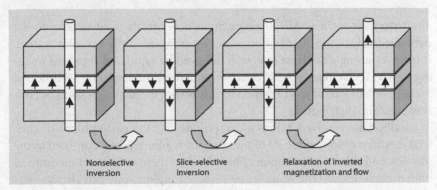

Nonselective inversion   Slice-selective inversion   Relaxation of inverted magnetization and flow

Fig. 46. Diagram of black blood MRA with double inversion recovery. The black arrows represent the longitudinal magnetization in the corresponding voxels

— The blood must be outside the scan plane during the two inverting pulses for its magnetization to be inverted.

— The blood must flow into the scan plane between double inversion and signal collection.

— Central k-space must be collected when the relaxing blood magnetization passes through zero. The interval between double inversion and the start of data collection is automatically calculated by the scanner's software.

Double inversion recovery can be combined with an additional inversion pulse to selectively rotate the longitudinal magnetization of fat into the negative z-direction prior to scanning. This will additionally suppress the signal from fat, as with a STIR sequence (▶ Chapter 7.5).

In the routine clinical setting, currently only 2D implementations of black blood MR angiography are available. The signal from slowly flowing blood as in the trabecular structures near the walls of the cardiac chambers may be difficult to suppress. The use of SE sequences makes the method somewhat slower than GRE-based techniques. Black blood MRA is an angiographic technique in the true sense of the word in that it primarily visualizes the vessel walls rather than the blood. The diagnostic accuracy of black blood angiography is not impaired by turbulent flow and the method may have a lower rate of false-negative results in the evaluation of atherosclerotic lesions, especially in patients with early disease before significant narrowing of the vessel lumen has occurred. However, typically only short vessel segments can be efficiently evaluated.

11.1.3 Time-Resolved MR Angiography

The term *time-resolved MR angiography* is now mostly used to refer to the dynamic study of the distribution of a contrast agent in the vascular system. It involves a single contrast injection followed by rapid and repeated imaging of a vascular region with contrast-enhanced angiography techniques. Successive MRA data sets obtained in this way map the progression of contrast medium distribution.

Ideally, time-resolved MR angiography depicts the early phases of contrast medium inflow, when all of the contrast medium is still confined to the arteries, and subsequent "venous phases" when there is contrast medium in both the arteries and the veins. Time-resolved angiography can also encompass evaluation of organ perfusion, as has been shown for the kidneys.

When time-resolved MRA images are updated fast enough, timing of the acquisition start is less critical for a successful differentiation of arteries and veins. Moreover, the method readily identifies the false lumen in dissections and facilitates the identification of retrograde contrast medium inflow. Finally, the time-resolved information enables detailed evaluation of the vascular system supplying and draining an arteriovenous malformation or a tumor.

The demands on scan time are even higher for time-resolved MRA compared with contrast-enhanced MRA. The minimization of scan time is a major concern but is usually achieved only at the cost of spatial resolution. In addition to the methods for scan time reduction already described, there are strategies that are specific for dynamic imaging. A widely used approach is the reconstruction of data sets for which the periphery of k-space has not been updated (TRICKS or time-resolved imaging of contrast kinetics, keyhole imaging). For image reconstruction, peripheral k-space data is taken from an earlier or later measurement and combined with more frequently updated central k-space data to form a complete raw data set. In the combined data sets, the data from the center of k-space encodes the most recent change in signal intensities. The three-dimensional k-space may be divided into different areas where the image information is updated at different intervals. Data is updated more frequently, the closer the area is to the center of k-space. In combination with the methods of reducing scan time discussed earlier such techniques permit the acquisition of 3D data sets in 1 to 6 sec, depending on the size of the imaging volume and the desired spatial resolution.

If even faster image acquisition is desired, one can dispense with phase encoding in the slice-select direction. In this way, one obtains two-dimensional images that represent projections of the signal intensities through the scan volume, similar to conventional X-ray techniques. Depending on the situation then, images can be updated several times per second with good spatial resolution.

11.2   Perfusion-Weighted Imaging

Perfusion-weighting MR acquisition techniques image tissues with signal intensities that vary in proportion to the flow of blood through their capillary beds. Thus, perfusion-weighted imaging (PWI) allows evaluation of microvascular circulation. It provides direct information on tissue perfusion, regardless of whether blood is supplied through the main vessel or collaterals. Perfusion imaging is mainly used to asses microvascular blood flow in the brain, the myocardium, the lungs, and the kidneys. In vivo, the degree of tissue perfusion is estimated from the magnitude of the signal change that is locally induced by a tracer entering the tissue of interest. Exogenous and endogenous tracers are distinguished.

An example of exogenous tracers are the gadolinium-based contrast agents used in contrast-enhanced MRA. These agents strongly affect the tissue signal when they flow into the target organ so that regional differences in perfusion are directly seen on the images (first-pass imaging).

The blood itself can be used as an endogenous tracer. To this end, the longitudinal magnetization of the blood in a feeding artery is saturated or inverted (arterial spin labeling, ASL). When the labeled blood arrives in the target anatomy before complete relaxation of its magnetization has occurred, it produces a decrease in signal. Because the signal decrease caused by the inflowing blood is usually too small to be seen directly, it is accentuated by means of subtraction of two data sets acquired with and without presaturation of the inflowing blood.

A paramagnetic contrast agent passing through a tissue induces transient shortening of its relaxation times, which is seen as an *increase in signal on T1-weighted images* and a *decrease on T2- or T2*-weighted images*. Both effects are exploited in MR imaging.

Contrast medium-based perfusion imaging of the heart, lungs, and kidneys is typically performed with T1-weighted GRE sequences. For cardiac perfusion imaging, the sequence must be synchronized with the cardiac cycle and generate at least one image from exactly the same phase of the cardiac cycle every second heartbeat. Sequences that collect more than one echo per excitation (multishot echo planar imaging, ▶ Chapter 8.5) are currently most often used and most widely investigated, due to their short scan time. Perfusion imaging of the lungs and kidneys is usually performed using T1-weighted 3D GRE sequences. Evaluation of contrast medium arrival in the target anatomy can be supplemented by monitoring the rate of contrast outflow from the tissue. Naturally, this involves longer scan times.

Cerebral perfusion imaging is more commonly done with T2*-weighted 2D or 3D echo planar sequences which depict the passage of the contrast medium as a transient decrease in signal intensity (dynamic susceptibility contrast-enhanced MR imaging). With these sequences, most of the brain can be imaged with acquisition of a new image stack every 1 to 2 seconds.

Ideally, one would determine absolute blood flow per unit time for each voxel of the target anatomy, e.g., given in milliliters per second and gram of tissue. In this way, one could not only identify small areas with reduced flow by a signal dynamics that differs from the one in their surroundings but also reliably diagnose globally reduced perfusion of an organ. Unfortunately, absolute quantification of blood flow is difficult to accomplish with exogenous as well as endogenous tracers although many published studies report absolute values. Numerous factors have to be taken into account with both techniques and a review of the most recent literature suggests that there is still no final agreement as to the most suitable approach, at least with regard to the contrast medium-based methods.

Given these problems with absolute quantification of blood flow, various parameters have been proposed to characterize the signal dynamics descriptively. Several of these parameters have been shown to correlate more or less consistently with an independently determined degree of tissue perfusion. Perfusion parameters used in dynamic contrast-enhanced MRI include, e.g., the time to peak signal enhancement measured from the moment the first change is observed or the relative signal change per unit time (enhancement slope). Although such evaluations yield numerical results that are largely observer-independent, they are nevertheless limited because they depend on the pulse sequence used and on other scan parameters. Hence, the perfusion parameters have to be calibrated after each change in the experimental setup and results from different study centers are difficult to compare.

Compared with various other modalities, MR perfusion techniques have the advantage of allowing noninvasive or minimally invasive evaluation of blood flow in a tissue with good spatial resolution. MR imaging involves no ionizing radiation exposure and is relatively fast. Patients can therefore be examined repeatedly, for example, to monitor therapy or to follow up surgery. Moreover, perfusion measurement can be performed in combination with other MR measurements in a single session. These can provide detailed anatomic information or help differentiate viable tissue regions with decreased perfusion from scar tissue or areas of acute infarction.

11.3 Diffusion-Weighted Imaging

Diffusion-weighting MR acquisition techniques image tissues with signal intensities that vary in proportion to the average distance by which water molecules are displaced per unit time through the process of water self-diffusion. The regional tissue or fluid signal in diffusion-weighted imaging (DWI) decreases with increasing average speed, with which water molecules travel in the direction of a magnetic field gradient that is temporarily generated by the MR scanner. The operator can control both the diffusion direction and the degree of diffusion-induced signal loss observed in a given acquisition by selecting the direction and the amplitude of the applied field gradient, respectively.

The magnitude of water molecule displacement is described quantitatively by the diffusion constant, which generally varies with the direction of diffusion.

When the net random microscopic translational motion of the water molecules results in an average distance of displacement that is the same in all directions the diffusion is called isotropic (▶ Fig. 47). In such a situation the diffusion-weighted MR images are independent of the direction along which the magnetic field gradient is applied. In the human body, nearly isotropic diffusion occurs in body fluids with freely mobile water molecules such as the cerebrospinal fluid (CSF) in the ventricles or cystic fluids. The diffusion constants and average molecule displacements in these fluids are rather large and are identical in all directions. This results in a strong signal attenuation on diffusion-weighted images.

In an environment that is structured at the microscopic level, the diffusion of water molecules is no longer unhindered but restricted by the geometry of the microscopic structures. Since the environments do not normally have a spherical geometry, the restriction of the water self-diffusion is different in different directions. The water self-diffusion is then directionally dependent or anisotropic. In the brain, for example, water molecules diffuse faster in the direction of axons with intact myelin sheaths than perpendicular to the axons. The diffusion constant is higher along the longitudinal axis of the axons than in the plane perpendicular to the axis. The diffusion-induced signal loss is larger when the diffusion gradient is applied in the direction along the axis of the fiber tract and smaller when it is applied perpendicular to the axis.

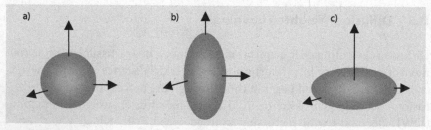

Fig. 47a–c. Diffusion tensor ellipsoids for isotropic (**a**), tubular (**b**), and layered environments (**c**)

Diffusion of water molecules in an anisotropic environment is restricted by structures that are smaller than the resolution of an MR image. Therefore, the directional differences can be observed only when most axons in a voxel are arranged in parallel.

Diffusion-weighted images depict lesions caused by stroke already within the first 6 hours of the onset of symptoms – before traditional MRI techniques such as T2-weighted images will show any significant changes. In the acute phase, the diffusion-induced loss of signal is less pronounced in affected areas and they appear brighter than unaffected regions. The positive contrast is gradually lost in the course of some days and finally becomes negative as a result of greater mobility of the water molecules.

Diffusion-weighted images are typically acquired with an echo planar imaging sequence. A pair of magnetic field-gradient pulses is delivered between the excitation pulse and signal collection to sensitize the sequence to diffusion effects (▶ Fig. 48). The pulse pair differs from that used in phase-contrast angiography in that both halves have the same polarity. However, the effect is very similar due to the 180° RF pulse which is delivered between both halves of the pulse. A change in phase is imparted to those spins that move along the field-gradient direction while the pulses are being applied. As a result, if spins in a voxel are displaced by different amounts in the process of diffusion, they all experience different phase shifts. Their magnetic moments do not add constructively any longer, which results in a weaker MR signal from the voxel. The signal attenuation depends on the strength and the duration of the gradient pulses, their temporal separation, and the diffusion constant along the direction of the gradient field.

The amount of diffusion weighting achieved with a given gradient pulse pair and inversion pulse sandwich is measured by the so-called b-value. It quantifies the amount of signal loss to be expected with a given pulse sequence for a given diffusion constant.

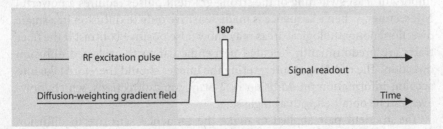

Fig. 48. Diagram of a diffusion-weighted sequence

The diffusion constants in biological tissues can be quantitatively measured by repeated scanning with different b-values but otherwise identical imaging parameters, in particular an unchanged gradient direction. The observed diffusion constants are referred to as *apparent diffusion coefficients* (ADC), to differentiate them from the constant of unrestricted diffusion in pure water.

Images whose gray-scale values represent the mean ADCs of the corresponding voxels are known as *ADC maps*. An area of acute infarction that is bright on a diffusion-weighted image (reduced mobility of the water molecules) will appear dark on the corresponding ADC map (smaller diffusion constant).

Diffusion constants for different directions can be measured by changing the direction of the gradient field. Such measurements provide detailed information on the local geometry of the microscopic structures that restrict water diffusion. This version of diffusion imaging is known as *diffusion tensor imaging* (DTI). For example, after measurement of the diffusion constants in six selected directions, the diffusion-restricting geometry can be described with the formalism of three-dimensional tensors. In this formalism, the diffusion-restricting geometry is approximated by an ellipsoid whose three main axes may differ in length in proportion to the ADC in the corresponding direction (▶ Fig. 47). Increasingly accurate geometric models of the averaged structures that hinder diffusion in a voxel may be generated when additional diffusion constants for other directions are measured.

The main current application of diffusion tensor imaging is tractography (fiber tracking) in the cerebral white matter. Here, it is attempted to reconstruct the spatial course of fiber tracts over longer distances by analyzing the relative orientation and size of diffusion ellipsoids in neighboring voxels.

Diffusion-weighted images are highly sensitive to all kinds of motion. These include rotation or trembling of the head in cerebral imaging or respiratory motion in imaging of the trunk. This is why short scan times are important. Fast switching of the strong gradient pulses requires a powerful MR scanner. When a sequence is made sensitive only to diffusion in a single direction, nonpathological areas may show false positive contrast if the fiber tracts are predominantly oriented perpendicular to the selected diffusion direction. The radiologist interpreting the images should therefore take into account information on diffusion in 3 orthogonal directions, which, however, can be obtained with a single scan.

The gradient pair applied to make the sequence sensitive to diffusion processes only attenuates the signal compared to images obtained without the gradient. Structures such as CSF with a strong signal on corresponding non-diffusion-weighted images may still appear bright on images with only mild to moderate diffusion weighting even though their diffusion constant is high. This effect is referred to as T2 shine-through and may be difficult to distinguish from actual restriction of diffusion. Only on strongly diffusion-weighted images are the signal intensities predominantly determined by diffusion.

Diffusion-weighted imaging is an area of intensive research because it provides unique information that cannot be obtained with other methods or only to a very limited extent.

11.4 The BOLD Effect in Functional Cerebral Imaging

Functional magnetic resonance imaging (fMRI) of the brain aims at identifying cerebral areas that respond to a well-defined external stimulus by a change in signal (brain mapping). Functional images are typically acquired using T2*-weighted techniques. Classical tasks used to induce neuronal responses are visual (such as looking at changing patterns) or sensorimotor (such as a sequence of defined finger movements) activation. A wide variety of protocols exist for neuronal activation and its synchronization with data acquisition (paradigms).

The experimentally observed signal variations in "activated" brain regions are assumed to be originally triggered by an increased oxygen demand in the activated region. To meet the higher demand, capillary blood flow and the blood volume in the activated region are increased by local vasodilatation. Moreover, it is assumed that the increased blood flow exceeds the metabolic needs after a physiological reaction time, which leads to an increased oxygen concentration in the blood of the local capillary bed. The higher oxygen loading of the blood's hemoglobin molecules prolongs the T2* time of the surrounding water, which is believed to explain the observed signal increase on T2*-weighted images. This contrast mechanism is known as blood oxygen level-dependent (BOLD) contrast.

While BOLD imaging is thought to reflect the oxygen concentration of blood, there are other functional MRI techniques that take advantage of the higher blood flow or the increased blood volume to localize cerebral activation.

BOLD imaging is typically performed with strongly T2*-weighted GRE EPI sequences (► Chapter 8.5) that allow scanning of the entire brain in a few seconds. The signal changes induced by activation are fairly small. Thus all slices are repeatedly imaged during alternating "on" and "off" periods of brain activation (block design paradigm, ▶ Fig. 49). Each voxel in the imaging volume is then assigned a statistical probability that variations of its signal intensity over time are caused by the on- and off-switching of brain activation. Voxels for which this probability exceeds a certain threshold value are color-coded on a so-called activation map. For the visual interpretation and localization of the activated regions the activation maps can be fused with traditional morphologic MR images that depict anatomic structures with a better spatial resolution.

The BOLD contrast increases with the magnetic field strength of the MR scanner. The noise that is associated with MR scanning makes it somewhat difficult to measure cerebral activation by auditory stimuli. The standard techniques of stimulation offer only a limited temporal resolution for the registration of physiologic changes. Here, event-related paradigms with only short periods of activation may be advantageous. The spatial resolution of BOLD imaging is fundamentally limited because the area with an increased oxygen saturation of the blood may be much larger than the region that has originally been activated. Finally, T2* is affected by many factors at the microscopic level that are difficult to isolate. It may prove impossible to relate the magnitude of the observed signal changes induced by stimulation to well-defined physiological parameters.

In relation to the extensive research activities focused on functional MR imaging it has a smaller role in routine clinical examination at most radiological centers. Clinically, BOLD imaging is for instance used to plan neurosurgical interventions. Despite its limitations, functional BOLD imaging

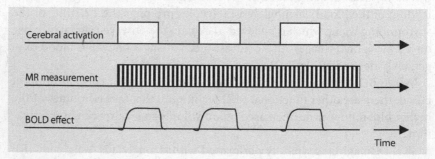

Fig. 49. Block design paradigm for functional brain imaging

enables fully noninvasive and radiation-free evaluation of subtle changes in cerebral activity with a spatial resolution of 1–2 mm or better and a temporal resolution in the range of 100 msec.

11.5   Cardiac Imaging

Imaging of the heart differs from imaging of other organs in that the constant cardiac motion causes blurring and other artifacts along the phase-encoding direction on MR images acquired with long scan times. With state-of-the-art equipment, though, the scan time for acquisition of a single slice can be reduced to such an extent that cardiac motion can be monitored on a series of images acquired in near-real time without detrimental degradation of image quality by artifacts caused by respiratory or cardiac motion. Most artifacts can be effectively eliminated when the scan time is less than 50 msec during systole and less than 200 msec during diastole.

> So far, real-time cardiac imaging has mainly been performed to rapidly localize the heart and long- and short-axis views for subsequent data acquisition.

To improve the spatial or temporal resolution of "real-time" cardiac imaging, the data acquisition for an image needs to be distributed over several heartbeats (so-called segmented acquisition). This is possible because cardiac motion is periodic under normal conditions and the myocardium will be at identical locations at identical phases of different cardiac cycles. To ensure that all data for an image is sampled during the same phase of the cardiac cycle, segmented data acquisition must be synchronized with the individual patient's heart rate. To this end, an electrocardiogram (ECG) is recorded and the data is used by the scanner software to identify the R wave in each cardiac cycle. The ECG data can be used in two ways, either to trigger MR acquisition to a specific phase of the cardiac cycle (cardiac triggering, prospective cardiac gating) or to retrospectively assign continuously acquired data to the corresponding cardiac phases (retrospective cardiac gating).

The scan time per cardiac cycle is shorter when an image acquisition is distributed over several cycles, which allows visualization of cardiac motion with improved temporal resolution. However, the overall scan time per image is longer and artifacts introduced by respiratory motion become more severe.

One possible solution is breath-held imaging. For instance, with a repetition time (TR) of 3.5 msec and collection of only one phase-encoding step per TR, 14 phase-encoding steps can be sampled in 50 msec. If we want to generate an image with a resolution of 224 pixels in the phase-encoding direction, the acquisition will have to be distributed over $224/14 = 16$ heart beats. However, patients with heart disease may find it difficult to hold their breath for 16 heart beats.

Since respiratory motion is also periodic, data acquisition cannot only be distributed over several cardiac cycles but also over several breaths. This is accomplished by monitoring the patient's breathing rhythm: a short 1D scan is alternated with image data acquisition for localizing the boundary between the diaphragm and the lung along the body's longitudinal axis. In this way, the image data can be – prospectively or retrospectively – associated with the different phases of the respiratory cycle (navigator technique). Using the navigator technique, scanning is not limited to the duration of a breath-hold but can be performed with the patient breathing freely. However, navigator techniques tend to be limited by inefficient data acquisition and long scan times. Moreover, they tend to provide the best results in healthy subjects with a fairly regular heart rate and breathing pattern.

When used in combination with these gating techniques, traditional MR imaging techniques already described allow the visualization of the anatomy of all cardiac chambers and of the vessels entering and leaving the heart, in three dimensions, without ionizing radiation exposure and with good SNR. MRI can thus be used to repeatedly examine patients with suspected congenital malformations, cardiomyopathies, valve incompetence, or pericardial disorders; to follow up patients after bypass surgery; and to monitor heart transplant recipients. For imaging of the coronary vessels, a wide variety of pulse sequences and sequence modifications are in use, which all have specific advantages and disadvantages. However, the major strengths of cardiac MR imaging in assessing coronary artery disease currently lie in the repeatable evaluation of morphology, function, and perfusion without radiation exposure as well as in the localization and precise delineation of infarcted areas.

Some specific applications are discussed in more detail below.

11.6   Cardiac Imaging with SSFP Sequences

Steady-state free precession imaging has become a fixed component of standard cardiac MRI protocols. In comparison to other GRE sequences SSFP sequences (► Chapter 7.7) yield images with a stronger blood signal despite shorter TR times (about 2–5 msec). It is thus possible to rapidly image the blood in the cardiac chambers with good contrast relative to the myocardium. Good contrast is maintained even when there is only little blood flow through the scan plane and the blood signal is not enhanced by inflow effects. This may be particularly advantageous in acquisitions of long-axis views of the left ventricle.

The sequence is usually acquired in the cine mode where each slice is imaged in multiple phases of the cardiac cycle. If, for instance, we again assume a single breath-hold acquisition with scan time segments of 50 msec and a patient with a heart rate of 70 beats per minute, myocardial wall motion could be observed on a sequence of 17 images from different phases of the cardiac cycle. After acquisition of several slices during different breath-hold periods, the motion of the entire heart can be evaluated and quantified. Even the apex of the heart is well accessible for evaluation on long-axis views. The option of a 3D acquisition of multiple slices during a single breath-hold period is now available on most modern scanners.

The acquired image data sets can be used to determine global morphologic and functional parameters such as the myocardial mass, the ejection fractions of both ventricles, or the stroke volumes. These parameters can be determined directly without having to make geometric assumptions as in the classical model-based methods. There is good interobserver reproducibility of the results.

Besides estimation of the global parameters, the method can also provide information on regional functional parameters such as local wall motion, left ventricular wall thickening from diastole to systole, or circumferential shortening. Disturbed perfusion can be diagnosed with a high degree of accuracy if a myocardial region showing normal wall motion at rest becomes hypokinetic during drug-induced stress (dobutamine).

A limited additional enhancement of the blood signal on SSFP images is observed in the presence of moderate concentrations of an MRA contrast agent.

The quality of SSFP images may be severely degraded by inhomogeneities in the static magnetic field, especially in connection with flow effects, as well as by an off-resonance setting of the RF frequency. However, the technical problems have been solved to such an extent that SSFP has become very reliable for routine clinical application.

11.7 Myocardial Perfusion Imaging

Myocardial perfusion is typically estimated from the signal enhancement seen on T1-weighted MR images that are obtained during the first pass of a contrast medium through the muscle tissue. Ideally, the images of all slices are updated once every heart beat. The contrast medium is injected intravenously, usually at a lower dose than administered for angiography. Ischemic areas are identified directly by a delayed inflow of contrast medium and/or a lower peak signal intensity during passage of the contrast medium. The differences to adjacent myocardium with normal perfusion are especially salient when viewing the images in rapid succession in the cine mode. In this way, it is also possible to identify disturbed perfusion confined to inner myocardial layers, which is more difficult to detect with competing diagnostic modalities.

Imaging is performed during drug-induced stress (adenosine, dipyridamole) and breath-hold. The actual scan takes less than a minute. Regions of reduced perfusion can be differentiated into viable and nonviable areas by combining stress imaging with a perfusion measurement at rest and late-enhancement imaging (► Chapter 11.8).

The most widely used techniques for myocardial perfusion imaging comprehend fast GRE or multishot EPI sequences that are preceded by a preparatory RF pulse. The preparatory pulse is either a 90° saturation pulse or a 180° inversion pulse, defining the sequence as a saturation recovery sequence or an inversion recovery sequence, respectively. The latter allows stronger T1 weighting while the former yields more stable signal intensities that are less sensitively affected by an irregular heart rate and, thus, better reproducible results. Depending on parameter settings and the specific options available on different scanners, it is currently possible to acquire up to about four slices per heart beat or eight slices every second heartbeat.

For quantitative analysis, the temporal course of the contrast medium concentration in the myocardium needs to be put in relation to the temporal course of the concentration in the inflowing blood. Since the latter cannot be measured directly for each voxel, the temporal course of signal intensity in the blood in the left ventricle is typically used as an approximation of the input function for all voxels. The quantitative analysis is associated with a number of problems: (i) different signal enhancement in myocardium and blood due to (a) the chosen set of pulse sequence parameters, (b) different local molar relaxivities of the contrast agent, or (c) non-quantifiable in-flow effects, (ii) unclear effect of water exchange through cellular and capillary walls, (iii) unknown permeabiltiy of the capillary membrane for the contrast medium, (iv) signal differences due to local variations in sensitivity of the receive coil, and (v) an ill-defined extent of signal enhancement resulting from respiratory motion of anatomic structures through the imaging plane. Various options are available to tackle each of these problems.

Despite the difficulties, results reported in the literature suggest that numerical parametric as well as quantitative data evaluations may be fairly independent of the examiner and reach a high diagnostic accuracy in comparison with several reference modalities.

11.8 Late-Enhancement Imaging

Late-enhancement images acquired about 10 to 20 minutes after intravenous administration of an angiographic contrast agent dose positively contrast myocardial regions which retain a higher contrast agent concentration than the surrounding normal myocardium. Solid evidence in the literature indicates that this allows the identification of both acutely infarcted tissue and scar tissue after an older infarction with good resolution ("bright is dead"), even though the contrasted area may be somewhat larger than the infarct at a very early stage. The increased late-phase contrast agent concentration in these areas is attributed to a larger extravascular, extracellular volume and/or slower washout. Validation studies performed so far suggest that late-enhancement images allow very accurate estimation of the size of an infarcted area.

However, late enhancement is not a specific feature of myocardial infarction. A similar signal enhancement may also be seen in myocardial regions affected by other heart diseases. While enhancement associated with infarction is usually confined to subendocardial regions with transmural extent in severe cases, the late enhancement seen in other disorders may be confined to the middle layer of the wall.

It must also be noted that very poorly perfused, nonviable areas may not show contrast enhancement due to failure of the contrast medium to enter these areas by the time the images are acquired. This applies especially to images that are obtained within the first minutes of contrast medium administration. Enhancement may be absent, e.g., in the central area of extended infarcted regions, while their periphery still appears bright. This phenomenon is described by such terms as "microvascular obstruction". It has been suggested that microvascular obstruction persisting for several days is associated with a particularly poor prognosis.

Late-enhancement images are acquired with GRE-based inversion recovery sequences. The recovery time between the RF inversion pulse and data acquisition (*TI*, inversion time) is selected such that the magnetization of healthy myocardium passes through zero when the central k-space lines are filled, leaving normal tissue dark on the resultant image. If imaging of the whole heart takes several minutes, it may become necessary to readjust TI to the changing contrast medium concentration. Late-enhancement imaging can be performed with 2D or 3D sequences.

While the differentiation of infarcted and healthy tissue has most often been straightforward, the differentiation of a subendocardial infarction and blood in the left ventricle may be difficult and require additional scans, for example with a different *TI*.

It has been suggested that late-enhancement imaging has the potential to become the method of first choice for demonstrating myocardial infarction and estimating its extent.

11.9 Detection of Increased Myocardial Iron Concentrations

MR imaging may have the potential to reliably detect excessively high iron concentrations in the myocardium on the basis of their T2*-shortening effect when precisely defined protocols are used for data acquisition and analysis. Such protocols can comprise acquisition of a short-axis slice through a central portion of the left ventricle which is repeated with different echo times using a GRE sequence. The signal loss observed with increasing echo times allows one to calculate the T2* relaxation constant in a region of interest, which may be placed in the septum. Preliminary results, mostly obtained in thalassemia patients, suggest that an extended range of reduced T2* values can be observed before a further T2*-reduction is associated with a deterioration of cardiac function. The T2*-threshold for impaired cardiac function was about 20 msec at 1.5 tesla compared with a mean value of 52 msec in healthy subjects. Measurement of T2* relaxation times might therefore provide the basis for identifying those patients who will benefit most from intensive iron-chelating therapy and so may be spared the poor prognosis associated with impairment of cardiac function.

12 MR Contrast Agents

JOHANNES M. FROEHLICH

Image contrast in medical MR imaging results from differences in signal intensity (SI) between two tissues and is determined by intrinsic and extrinsic factors. These are respectively properties of the different tissues and properties of the MR scanner, especially of the pulse sequence used.

MR contrast media are pharmaceutical preparations that are administered in MR imaging to further enhance the natural contrast and additionally to obtain dynamic (pharmacokinetic) information. To achieve these goals, contrast agents used for MRI must have specific physicochemical properties and also a suitable pharmacokinetic profile.

MR contrast media fundamentally alter the intrinsic contrast properties of biological tissues in two ways:
— *directly* by changing the proton density of a tissue or
— *indirectly* by changing the local magnetic field or the resonance properties of a tissue and hence its T1 and/or T2 values.

The local magnetic field strength is altered because the unpaired electron spins of the contrast medium (CM) interact with the surrounding hydrogen nuclei of the water, fat, or protein molecules in the tissue. Thus, the mechanism of action of an MR contrast agent comprises processes of the electron shell and not just processes at the nuclear level, as does the MR effect. The magnetic moments of electrons are 657 times greater than those of protons. This is one of the reasons why the electron shell has much more powerful paramagnetic properties than a hydrogen nucleus.

The interactions occurring between contrast medium electrons and tissue protons comprise "inner-sphere relaxation" (through interaction with bound water) and "outer-sphere relaxation" (e.g. arising from the diffusion of water nearby). Both processes contribute substantially to the overall effect of MR contrast media.

Before we proceed with our discussion of contrast media, some terminology must be clarified:

— *Paramagnetic substances* have magnetic moment (resulting from individual spins) because they consist of atoms or molecules that have magnetic moment due to unpaired electron orbits in their outer electron shells or unpaired nucleons in their atomic nuclei. When such a substance is exposed to an external magnetic field, most of the magnetic moments align with the direction of the magnetic field and the magnetic moments add together, resulting in a local increase in the magnetic field (just as with protons). When no external magnetic field is present, the magnetic moments occur in random pattern and there is no net magnetization. Many dissolved metal ions (including iron in blood) but also stable radicals are paramagnetic because they contain unpaired electrons. Examples are Co^{2+}, Co^{3+}, Fe^{2+}, Fe^{3+}, Gd^{3+}, Mn^{2+}, Mn^{3+}, and Ni^{3+}. Because of their powerful magnetic moment (see above), substances with unpaired electrons are preferred as MR contrast media. Most of the clinically available MR contrast media are paramagnetic metal ion compounds (gadolinium chelates, manganese, iron).

— *Superparamagnetic substances* have very strong paramagnetic properties. Their greatly increased magnetic moment (10- to 1000-fold) results from the arrangement of the paramagnetic ions in a rigid crystal lattice, which increases the mobility of their surrounding electrons (e.g. iron oxide in the form of superparamagnetic nanoparticles). Superparamagnetic contrast agents are solid substances that not only have T1 and T2 effects but also markedly distort the local magnetic field (magnetic susceptibility).

— *Ferromagnetic substances* also consist of large groups of atoms whose unpaired electron spins are strongly entwined by exchange coupling (solid state). These substances retain their magnetization even after the external magnetic field has been removed and subsequently become permanent magnets. The best known example is iron (Fe).

By far the majority of all substances are *diamagnetic* (strictly speaking, diamagnetism is present along with the other forms of magnetic properties in these substances). When brought into an external magnetic field, diamagnetic substances induce very weak overall magnetization in the opposite direction (-z) – mostly because the orbital movement of most electrons is counterclockwise.

Now we can address the question as to how a contrast medium alters the MR signal and thus enhances contrast on the resultant image.

Unlike radiographic contrast media, which are directly seen on an X-ray absorption image, an MR contrast medium, such as a gadolinium complex, acts indirectly by altering the relaxation properties of surrounding hydrogen protons. In the example shown (► Fig. 50), the image obtained after IV administration of a Gd complex (right) depicts two lesions which were not visible on the image acquired before contrast medium administration (left). To produce this effect, the contrast medium must have two properties – first, it must diffuse through the blood-brain barrier and, second, it must be able to interact with the local protons (thereby shortening their T1).

MR contrast agents can alter an MR image in one of four ways.

Changing Spin or Proton Density

The presence of a contrast medium affects the amount of protons present in a voxel. Most agents reduce the number (e.g. freon-like compounds such as perfluoro-octyl bromide [PFOB], barium sulfate, fatty emulsions). The decrease in local proton density is associated with a signal loss, as for example after oral administration of a barium sulfate suspension.

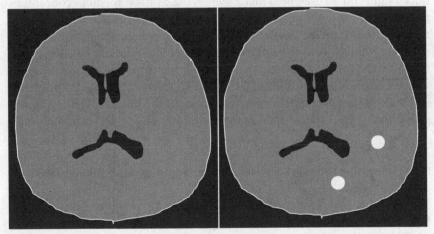

Fig. 50. Schematic representation of the CNS in transverse orientation. T1-weighted MR images before contrast medium administration (left) and after IV administration of 0.1 mmol Gd/kg body weight (right). Two additional lesions are seen on the contrast-enhanced image. The signal increase results from local T1 shortening and the extravascular distribution of the contrast medium within the lesions

Shortening T1 and T2 Relaxation Times (Most Important)

An MR contrast medium can be thought of as a catalyst that accelerates the relaxation of nearby protons by withdrawing the excess energy (in T1 weighting) the protons have previously absorbed from the excitation pulse (spin-lattice interaction). The faster *recovery* of longitudinal magnetization results in a stronger MR signal. An agent that enhances the signal is called a *positive contrast medium*. In addition, the magnetic moments of unpaired electrons alter the local magnetic field strength, resulting in faster dephasing due to spin-spin effects and enhancement of T2 relaxation. High contrast medium concentrations, e.g. in the lower urinary tract, produce local field inhomogeneities and T2 shortening, which is seen as a signal loss especially on T2-weighted images.

The significance of faster relaxation is nicely illustrated by comparing relaxation times in different environments:
— spontaneous relaxation in a vacuum: 1016 years,
— relaxation in a watery solution: about 1 second,
— relaxation in a watery contrast medium solution: a few milliseconds.

Pathologic tissue that takes up the contrast medium shows a signal change (typically a markedly higher signal on T1-weighted images and a reduced signal on T2-weighted images when the contrast medium concentration is high) while the surrounding normal tissue not containing any contrast medium remains unaffected. For optimal visualization of the effects resulting from the selective uptake of the agent into a lesion, it is important to adjust the imaging parameters, especially the weighting (e.g. predominantly T1 weighting and a short TR when a Gd preparation is administered) (► Fig. 51). Members of this class that are used in clinical MR imaging are gadolinium-based compounds, manganese compounds, and iron solutions.

Faster Dephasing Through Local Field Inhomogeneities (Susceptibility Effects)

So-called T2* effects are predominantly seen on T2-weighted images. Local field inhomogeneities caused by the high magnetic moment of the contrast medium accelerate dephasing of the protons beyond normal FID and thus shorten T2 further. This phenomenon is known as magnetic susceptibility and predominantly occurs in the presence of high local field strengths or at interfaces and may also cause artifacts. Susceptibility becomes manifest as a pronounced signal loss that is best appreciated on T2-weighted images. The

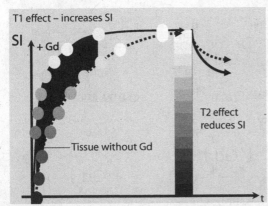

Fig. 51. Relaxation-time curve of a tissue with and without Gd uptake. The T1-shortening effect and resulting increase in signal intensity (SI) are transient (black area). The signal increase can be appreciated on images acquired with T1 weighting and a short TR. No SI increase will be seen when a longer TR is used (except as a result of the increasing T2 effect that occurs in the presence of high contrast medium concentrations)

agents producing a signal loss are therefore termed *negative contrast media*. Examples are superparamagnetic iron oxide nanoparticles (SPIO) that are taken up by the reticuloendothelial system (RES) of normal liver tissue and can thus serve as a liver-specific contrast medium for the selective suppression of the signal from normal liver.

Shifting the Resonance Frequency (Dysprosium)

Another mechanism of action is shifting of the resonance frequency by several hundred ppm. This effect is similar to chemical shift and weakens the measured proton signal. Dysprosium-based compounds are known to have this kind of effect but have virtually no role in medical MR imaging.

12.1 Chemical Structure

Most of the paramagnetic substances that can be used as contrast media are toxic metal ions with an unfavorable distribution in the body. This applies especially to gadolinium, which belongs to the lanthanide series of rare-earth elements. These substances may not be introduced into the body in their native state but only after having been chelated to a ligand. The ligands used for complexing should have a strong and specific affinity for the active ingredient (DTPA, DOTA, DTPA-BMA, HP-DO3A, BT-DO3A, BOPTA) (▶ Fig. 52). However, as complex binding is a reversible process (equilib-

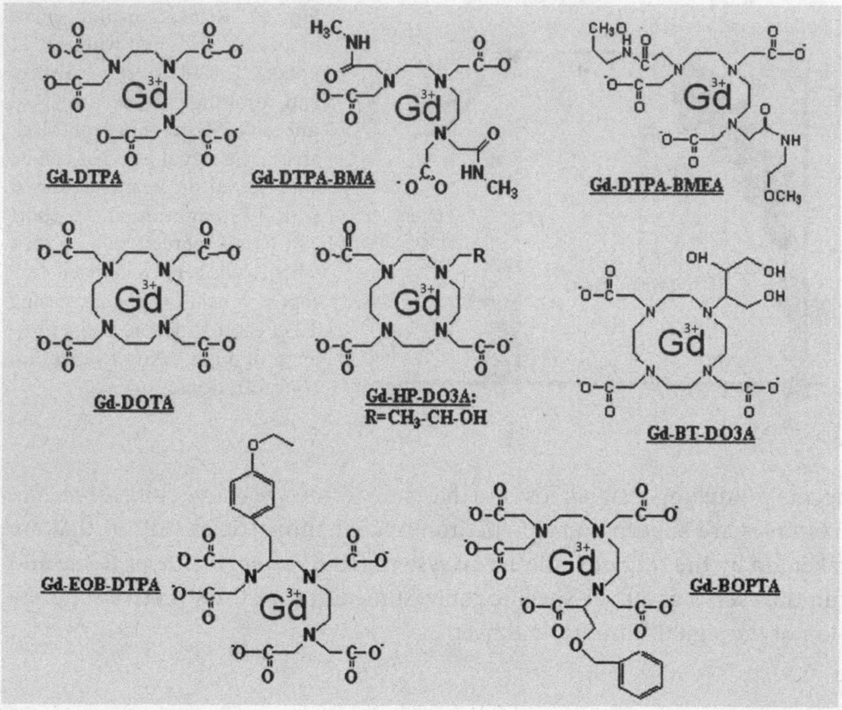

Fig. 52. Chemical structures of gadolinium-based compounds (linear structure: Gd-DTPA, Gd-DTPA-BMA, Gd-DTPA-BMEA, Gd-BOPTA, Gd-EOB-DTPA; macrocyclic structure: Gd-DOTA, Gd-HP-DO3A, Gd-BT-DO3A). Gd-BOPTA and Gd-EOB-DTPA have a more lipophilic side chain with a benzyl ring. These two compounds bind reversibly to protein. The side chain is also responsible for specific uptake into hepatocytes and partial hepatobiliary elimination (► Chapter 12.3.3). The other Gd-based compounds are unspecific agents that are eliminated via the kidneys

rium reaction between free and bound forms), a small portion of the central atom (most often Gd) may be released from the compound. The amount varies with the pH, temperature, and the presence of competing substances (for example other metal ions such as Cu^{2+}, Ca^{2+}, Zn^{2+}, $Fe^{2/3+}$ or acidic protons in the stomach) but is so small that no appreciable toxic effects occur. As an additional safeguard, most commercially available contrast medium preparations contain excess amounts of free complexes (typically Ca/Na complexes) to immediately intercept any gadolinium ions which are released

The toxicity of the gadolinium ions results from the fact that their diameter is virtually identical to that of calcium ions. Free gadolinium ions can thus function like calcium antagonists and block calcium channels by binding to them. This may impair cellular respiration, muscle contractility, and blood clotting. Besides greatly reducing the toxicity of gadolinium, the ligands determine the biodistribution of the compound. Unspecific, liver-specific, and macromolecular gadolinium-based compounds can be distinguished.

12.2 Relaxivity

Relaxivity is a measure of the relaxation efficiency of an MR contrast agent. It varies with the Larmor frequency and temperature but also with the concentration of the paramagnetic contrast medium preparation and properties of its molecular structure (such as the ability of the chelated ion to interact with water molecules, movement of side chains, or magnetic induction). By virtue of its seven unpaired electrons, trivalent gadolinium (Gd^{3+}) is one of the most powerful paramagnetic elements. So-called molar relaxivity is determined by measuring T1 or T2 in a one-molar solution that is obtained by dissolving 1 mole of the substance in 1 liter of water.

> Relaxivity: $R1 = 1/T1$ and $R2 = 1/T2$
> (concentration: 1 mol/liter, measured at a temperature of 20°C and a given Larmor frequency/field strength)

The higher the relaxivity, the better the interaction of the contrast medium with nearby water protons. This results in faster relaxation of the protons and an increase in signal (e.g. on T1-weighted images). When a substance has a high relaxivity, it is theoretically conceivable to reduce the dose as there is a direct relationship between the dose of a contrast agent and its relaxation enhancement effect. In the future there will be contrast agent preparations that can be applied either to reduce the gadolinium dose (to cut cost) or, contrarily, to increase the dose in order to achieve more pronounced signal enhancement, thereby improving contrast and enabling faster imaging.

The unspecific gadolinium-based compounds that have been available so far differ slightly in their contrast enhancement effects (and hence are administered at different dosages). Most of the commercially available prepa-

Table 7. R1 and R2 relaxivities of selected MR contrast media measured in water at a field strength of 1.0 T. Gadolinium chloride has very high relaxivities but is unsuitable as a contrast medium because it is highly toxic. Complexing reduces relaxation efficiency because it increases the distance between Gd^{3+} and nearby water protons

	R1	R2
$GdCl_3$	9.1	10.3
Gd-DTPA = Magnevist®	3.4	3.8
Gd-DOTA = Dotarem®, Artirem®	3.4	4.3
Gd-DTPA-BMA = Omniscan®	3.9	5.1
Gd-HP-DO3A = Prohance®	3.7	4.8 (at 0.5 T)
Gd-BT-DO3A = Gadovist®	3.6	4.1 (at 0.47 T)
Gd-BOPTA = Multihance®	4.6	6.2
Mn-DPDP = Teslascan®	2.3	4.0
Ferumoxides = Endorem® (SPIO)	40	160
Ferucarbotran = Resovist® (SPIO)	25.4	151
Ferumoxtran = Sinerem® (USPIO)	21.6	44.1

rations have a concentration of 0.5 mol/l, corresponding to an amount of gadolinium and dosage of 0.2 ml/kg body weight (= 0.1 mmol Gd/kg bw).

With the recent advent of new gadolinium-based CM preparations and new applications, it has become necessary to carefully choose the proper solution in terms of its specific physiologic interactions and concentration in order to achieve the desired relaxation enhancement effect for the intended purpose (► Table 7):

— In *direct MR arthrography* gadolinium solutions are administered at a dilution of 1:100 to 1:500, corresponding to a concentration of 5 mM Gd/l (= 0.005 mol/l) to 1 mM Gd/l (= 0.001 mol/l). A 2.0–2.5 mM Gd solution has proved to be most efficient. Intra-articular injection of an undiluted solution (containing 0.5 mol/l) would cause a signal loss instead of the desired enhancement. Conversely, IV injection of the dilutions used for arthrography would not produce enough signal and adequate contrast enhancement for lesion detection.

— *Gadovist®1.0* (gadobutrol) contains twice the amount of the active ingredient (1.0 mol of Gd per liter instead of the usual concentration of

0.5 mol/l). Using this preparation, one can reduce the dose in milliliters or the rate of administration. Thus, the recommended dose is 0.1 ml/kg body weight instead of the usual dose of 0.2 ml/kg bw. In this preparation, twice the concentration of the active ingredient is feasible due to the high water solubility of the gadolinium complex. This solution is expected to provide higher first-pass concentrations in certain vascular territories such as the peripheral arteries. However, as with higher concentrations of iodine-based contrast media, one also has to take into account the higher osmolality and viscosity of the solution.

— *Albumin-binding agents:* Contrast media like Gd-BOPTA (Multihance® – gadobenate dimeglumine), Gd-EOB-DTPA (Primovist® – gadoxetate disodium – marketed as a 0.25 mol/l preparation), or MS-325 EPIX (gadofosvesete trisodium – Vasovist®) bind reversibly to human albumin and therefore have a higher relaxivity in blood compared with water. The larger molecule size alters the contrast medium distribution and the enhancement effect varies with the protein affinity and the local protein concentration. How this affects dosing regimens, remains to be determined (► Fig. 52).

Finally, one has to take into account that both R1 and R2 can influence the MR signal.

The dose-effect curves of MR contrast media differ from the linear curves of X-ray contrast media in that they have a peak that indicates optimal contrast medium concentrations (► Fig. 53). The concentration which produces optimal enhancement varies with the MR contrast medium and the pulse sequence used. In general, the effects of lower concentrations are seen more clearly on T1-weighted images while T2 effects or decreases in signal intensity become more prominent at higher concentrations. From a practical perspective, this means that it is not necessarily the highest concentration or amount of gadolinium that produces the highest signal increase and hence the best contrast. Doubling the relaxivity does not imply doubling the signal intensity!

Relaxation enhancement also varies with field strength and tends to decrease as field strength increases (as T2 effects become stronger) while the contrast-to-noise ratios at 3.0 T are, in general, still superior to those achieved at 1.5 T or lower field strengths. Moreover, there may be variations among tissues resulting from differences in their intrinsic T1 and T2 times. As a rule, higher contrast agent doses are needed on scanners with low field strengths (0.2–0.6 T).

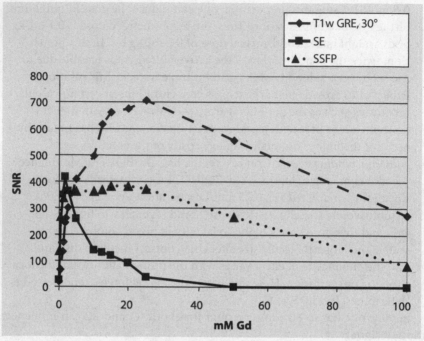

Fig. 53. Dose-effect curves of SNR after administration of an extracellular Gd-based contrast medium (Gd-DOTA) for three different pulse sequences at 1.5 tesla: SE = spin echo sequence, GRE = gradient echo sequence, SSFP = steady-state free precession sequence. Note the clearly different peak maxima

12.3 Pharmacologic Properties

The pharmacologic properties of a contrast agent, and especially its pharmacokinetics, determine its distribution in the body and hence its effect on the MR signal. Based on these properties, different types of contrast media can be distinguished (▶ Table 9).

12.3.1 Extracellular Contrast Agents

Extracellular contrast media are low-molecular-weight, water-soluble compounds that distribute in the vascular and interstitial spaces following IV administration. Most MR contrast media used today belong to this group of gadolinium(III) complexes (▶ Fig. 52). They are:

— Gd-DTPA (gadopentetate dimeglumine = Magnevist®/linear ionic complex),
— Gd-DOTA (gadoterate meglumine = Dotarem®/macrocyclic ionic complex),
— Gd-DTPA-BMA (gadodiamide = Omniscan®/linear nonionic complex),
— Gd-HP-DO3A (gadoteridol = Prohance®/macrocyclic nonionic complex),
— Gd-DTPA-BMEA (gadoversetamide = Optimark®/nonionic linear complex),
— Gd-BT-DO3A (gadobutrol = Gadovist®/linear nonionic complex),
— Gd-BOPTA (gadobenate dimeglumine = Multihance®/linear ionic complex; also used as a liver-specific agent).

Intravascular administration of a standard dose of an extracellular contrast medium shortens T1, producing an increase in signal intensity in the vessels (first pass, ▶ Chapter 11.1.1) and in the tissues due to tissue perfusion or disruption of the capillary barrier (brain, spinal cord, eyes, testes). Under normal conditions, these contrast media do not cross the blood-brain barrier because they are strongly hydrophilic (i.e. have a high affinity for water). The change in contrast medium distribution observed when the barrier is disrupted is an important diagnostic criterion. In general, the contrast effect is best appreciated on T1-weighted images, preferably in conjunction with fat suppression. The effect is comparable to that of water-soluble X-ray contrast media and is characterized by rapid diffusion of the contrast medium into tissue, thus equalizing the concentration between the vascular and interstitial spaces.

Extracellular contrast agents are eliminated renally by passive glomerular filtration. In this way, virtually all of the substance is eliminated unchanged without being metabolized. The plasma elimination half-life is about 90 min. Under normal conditions, far more than 90% of the administered dose is eliminated via the kidneys within 24 hours. The concentration in the kidneys produces additional T2 shortening, which may be seen as a loss of signal in the lower urinary tract. Only small amounts of contrast medium cross the placenta or are excreted in the breast milk. Current guidelines of the European Society of Urogenital Radiology (ESUR) no longer recommend that lactating women suspend breast-feeding after administration of a gadolinium contrast medium.

Extracellular contrast media are administered intravenously as a bolus or drip infusion (see also MR angiography, ▶ Chapter 11.1) at a *dose* of *0.1–0.3 mmol/kg* body weight. Higher doses of up to 0.5 mmol/kg bw have been administered sequentially in MR angiography. Since the majority of preparations are formulated as 0.5 molar solutions, the standard dose for single administration is 0.2 ml/kg bw (but only 0.1 ml/kg bw for a 1.0 molar formulation!). Some investigators recommend slightly higher doses for imaging at lower field strengths (<0.5 T) in order to achieve similar contrast enhancement to that on high-field MR scanners (1.5 T, 3.0 T).

From a practical perspective, it is important to note that the distribution half-life of about 2.5–5.0 min allows delayed imaging after IV administration, which is useful to assess extravascular pathology (such as tumors, metastases, or lesions) when enough contrast medium has reached the extravascular space. This does not apply to examinations in which evaluation of the early arterial or vascular phase is relevant (such as dynamic studies of the liver, pituitary gland, breast, and other organs).

In rare cases is imaging performed with a triple dose of contrast medium (0.3 mmol Gd/kg bw), which is actually a single dose followed by a double dose within 30 min of the first injection. It has been shown that this dosing regimen improves the detection of cerebral lesions in individual cases. The diagnostic gain is clinically significant, however, only in cases where it leads to therapeutic consequences (e.g. operable lesions, treatment of multiple sclerosis). Another option to improve the contrast medium effect is magnetization transfer contrast (MTC) imaging (▸ Chapter 3.6).

The adverse events that may occur after administration of an MR contrast medium are the same as for nonionic iodinated contrast media although adverse reactions are much less common because lower CM doses are required for MRI. Mild adverse events such as heat sensation, headache, nausea, or mild pseudoallergic reactions of the skin and mucosa occur in 1–2% of cases. Extravasated contrast medium can cause local pain and inflammatory reactions including tissue necrosis. Patients on asthma medication or with a history of contrast medium allergy are at an increased risk of allergoid reactions. Anaphylactoid shock induced by an MR contrast agent is extremely rare (about 1:50,000 cases). There is controversy about the renal tolerance of gadolinium preparations. Anecdotal cases of nephrotoxic effects after gadolinium administration have been reported. When identical volumes are administered, gadolinium preparations appear to be less nephrotoxic than X-ray contrast media but this is no longer the case when absolute amounts of the substances (molarities) are compared. As with other contrast agents, special caution is indicated when high doses are administered or in high-risk patients because elimination is much slower in these cases. In patients with end-stage renal failure, unspecific gadolinium chelates can be removed by dialysis.

Extracellular contrast media can be administered as a bolus, allowing them to be used for dynamic studies in conjunction with fast scan techniques, for instance, in contrast-enhanced MRA and liver imaging. Images obtained approximately 30 sec after IV administration show arterial anatomy and perfusion. About 1 min after contrast administration, the parenchyma is depicted (referred to as the portal venous phase in liver imaging). After about 3 min the distribution of the contrast medium in the extracellular (vascular and interstitial) space can be evaluated. Late enhancement images are acquired to evaluate specific washout phenomena (e.g. delayed washout in myocardial infarction after 10–20 min, ► Chapter 11.8). In contrast-enhanced MR angiography, arterial first-pass contrast medium dynamics can be evaluated.

Being heavy metal solutions, gadolinium-based contrast agents are radiopaque. However, because of the lower metal concentration, their X-ray absorption is only about one third that of the water-soluble iodinated contrast media (and their optimal kV value is different). A gadolinium-based contrast agent can be used for conventional radiography in patients with contraindications to iodine preparations (e.g. active thyroid disease) but only after carefully weighing the expected benefits against the agent's osmotic load, lower radiopacity, and different pharmacologic properties as well as higher price.

12.3.2   Intravascular or Blood Pool Contrast Agents

Intravascular or blood pool contrast media are higher-molecular-weight compounds with longer intravascular residence times due to the fact that they cannot diffuse through the capillary walls, or only very slowly, as a result of their molecule size. Some blood pool contrast agents have higher molar relaxivities since longer side chains in the ligand reduce Brownian molecular motion, thereby increasing the central atom's accessibility to water. The residence time in the vascular compartment and hence the imaging window varies with the molecular weight and elimination rate of the agent. However, when the capillary barrier is disrupted, leakage of blood pool contrast medium into the extravascular space provides information on the permeability of a lesion or damage to capillary membranes (tumor, trauma, hemorrhage, infection, irradiation). More recently, it has been shown that blood pool contrast agents appear to have a potential to detect occult gastrointestinal bleeding or to identify and characterize tissues with impaired capillary permeability such as tumors. Moreover, blood pool agents with their fairly constant intravascular concentrations (steady state) are expected to improve quantitative perfusion measurement as well. Vascular imaging using blood pool contrast media is not restricted to assessment of the arterial phase. However, overlapping veins may obscure arterial anatomy.

The following intravascular contrast agents are distinguished on the basis of their pharmacologic properties:

— Gadolinium or iron oxide micelles, liposomes, or nanoparticles (SPIO, USPIO, ► Chapter 12.3.5). These preparations have long circulation times due to their particulate nature. Most of the administered dose is inactivated by the reticuloendothelial system (RES). Various iron oxide nanoparticle preparations are currently being developed. These include Sinerem®/Combidex® and Supravist®.
— Macromolecular agents such as Gd-based dextrans or polylysines (gadomelitol = Vistarem®; Gd-DTPA cascade polymer = Gadomer-17). Macromolecular Gd chelates have higher relaxivities and have the advantage of being cleared by the kidneys.
— Albumin-binding low-molecular-weight Gd complexes have a lipophilic side chain (► Fig. 52) that enables their reversible binding to human proteins, thereby slowing down extravascular diffusion.

12.3.3 Liver-Specific Contrast Agents

Liver-specific contrast media are administered intravenously and accumulate in normal liver cells through anion receptor-mediated endocytosis but not in metastases or other tissues foreign to the liver (► Fig. 54). These agents are highly lipophilic Gd(III) or Mn(II) complexes. The route of elimination is biliary (enterohepatic circulation) and renal. A liver-specific contrast medium in clinical use is Mn-DPDP (mangafodipir trisodium – Teslascan® 0.01 mol/l solution – and 0.05 mol/l in the USA). This agent also accumulates in other organs such as the pancreas because manganese is released from the complex and the latter is metabolized. Gd-BOPTA or gadobenate dimeglumine (Multihance® 0.5 mol/l) accumulates in the liver parenchyma after 30–60 min following an initial phase that is rather unspecific. This preparation has been approved for the detection of focal liver lesions in some countries. Only 2-7% of the administered agent is excreted in the bile. The fact that

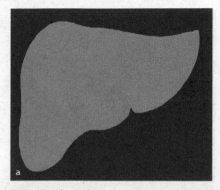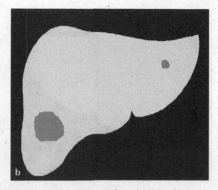

Fig. 54a, b. Schematic representation of the liver without (**a**) and with (**b**) administration of liver-specific contrast medium. Liver-specific agents such as Mn-DPDP, Gd-
BOPTA, and Gd-EOB-DTPA produce a diffuse increase in the SI of normal liver tissue
on T1-weighted images (**b**) compared with an unenhanced image (**a**). Lesions without
intact liver cells such as metastases thus become visible as negative or low-signal-intensity areas. This is why the contrast-enhanced image is a functional image. Liver-specific
contrast agents improve not only the detection of focal liver lesions but also their characterization

this contrast medium preparation acts both as an unspecific gadolinium
agent with the capacity to alter plasma relaxivity due to its predominantly
intravascular distribution as a result of reversible albumin binding and as
a liver-specific agent in the subsequent liver phase opens up new fields of
application in MR angiography, imaging of cerebral lesions, and in the detection of breast cancer and liver metastases. Gd-EOB-DTPA or gadoxetate
disodium (Primovist® 0.25 mol/l) has the highest specific absorption rate
with about 50% hepatobiliary elimination and has recently been approved
for the detection and characterization of focal liver lesions in Europe. It enhances normal liver tissue about 10–20 minutes after IV administration. Lesions not taking up the contrast medium thus show negative contrast and
are delineated as low-signal-intensity areas against the bright background of
normal liver tissue on T1-weighted images. Only in patients with biliary obstruction will there be little or no uptake of the contrast medium into liver
cells. Apart from improving detection of focal liver lesions, Gd-EOB-DTPA
may also have a potential for use in MRCP imaging, where it provides positive contrast of the bile ducts due to its biliary elimination. The manufac-
turer recommends the preparation for combined imaging of the dynamic
phase (first pass) and the hepatocellular late phase although this application
must be further validated.

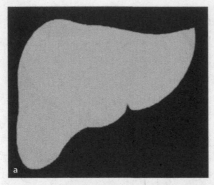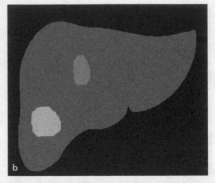

Fig. 55a, b. Schematic representation of the liver without (**a**) and with administration of a negative MR contrast medium (SPIO, **b**). RES-directed MR agents such as SPIO produce a diffuse decrease in the SI of normal liver tissue on short T2-weighted sequences (**b**) compared with an unenhanced image (**a**). Thus, lesions without Kupffer cells such as liver metastases are depicted with a high signal intensity (**b**, brighter lesion). Benign lesions often show reduced RES activity and therefore appear brighter than normal liver parenchyma but darker than on an unenhanced image (**b**, darker lesion). SPIO-enhanced images are functional images because they reflect RES activity. SPIO particles improve the detection of liver lesions as well as their characterization

12.3.4 RES Contrast Agents

MR contrast agents targeted to the reticuloendothelial system (RES) are administered intravenously and are predominantly phagocytosed by RES cells, in particular by Kupffer cells in the liver, and, to a lesser extent, in the spleen and the bone marrow. RES-specific agents used clinically are ferumoxides (AMI-25/Endorem®/Feridex®) and ferucarbotran (SHU 555 A/Resovist®). These superparamagnetic agents markedly shorten T2 and consequently reduce the SI of normal liver tissue with intact RES while neoplastic tissue without RES retains its bright signal (▶ Fig. 55). The rate at which the contrast medium is accumulated into a lesion can also be exploited diagnostically to improve lesion characterization since benign liver lesions such as adenoma, FNH, or hemangioma also have RES activity. Optimal contrast is achieved when using an intermediately T2-weighted sequence with a longer TR and a TE which should not be too long – that is, a sequence with a high susceptibility effect without an unduly long scan time.

The increase in intravascular signal associated with the T1-shortening effect can be used diagnostically to evaluate the vascularization of liver lesions such as hemangiomas.

Ferumoxides and ferucarbotran consist of iron oxide nanoparticles of different particle size that are coated with dextran and carboxydextran, respectively. They are administered as a bolus or as an infusion at doses of 8–15 µmol Fe/kg body weight. The optimal imaging window for exploiting the T2* effect is about 15 minutes to 8 hours after administration. These iron oxide preparations remain visible in the liver for 3–7 days. Thereafter, the metabolized iron enters the normal body iron metabolism cycle. A rare side effect whose mechanism is still unclear is back pain. Other rare adverse events are pseudoallergic reactions.

12.3.5 Lymph Node-Specific Contrast Agents

Contrast media specific to the lymph nodes have been in clinical trials for some time. They are superparamagnetic iron oxide nanoparticles (AMI-227, ferumoxtran, Sinerem®/Combidex®) that can be administered indirectly (subcutaneously), directly (endolymphatically), or intravenously. Other names used for lymphatic MR contrast agents are ultrasmall superparamagnetic iron oxide particles (USPIO) or monocrystalline iron oxide nanoparticles (MION). Following intravenous infusion, these agents remain in the blood for 24–36 hours before they accumulate in the lymph nodes and lymphatic vessels. As they are phagocytosed by macrophages and hence reach high local concentrations, they have a pronounced T2-shortening effect which decreases the signal of normal lymph nodes. Metastatic lymph nodes are thus identified because they do not take up contrast. Iron oxide nanoparticle preparations are also being developed as vascular contrast media (blood pool agents).

12.3.6 Tumor-Targeted Contrast Agents

MR contrast agents targeted to tumor cells are compounds such as metal porphyrins that accumulate in rapidly dividing cells. The mechanism of uptake is still unclear. These agents can be used to detect primary and secondary tumors or inflammatory tissue as well as to simultaneously perform photodynamic laser therapy (PDT). For therapeutic purposes, the accumulated metal porphyrins are activated to destroy surrounding tumor tissue by application of a high-energy beam. Unfortunately, tumor-specific contrast agents have a high systemic toxicity and have so far only been investigated in animals.

12.3.7 Other Emerging Tissue-Specific Contrast Agents

A number of different strategies have been devised to target contrast agents to specific tissues. Potential targets include not only tissue-specific antigens or epitopes but also genetic and functional features that distinguish tissues at the molecular level. Such specific MR contrast media consist of a paramagnetic or superparamagnetic signal emitter, a carrier structure (spacer), and a targeting system (monoclonal antibody, polysaccharide coat, coordination site for enzymes). After attachment of the contrast agent to a target, enzymes release a binding site or alter relaxivity, which is visualized on the image (*activation*). A similar mechanism underlies the identification of blood clots with a fibrin-binding experimental preparation, EPIX 2104R. Another promising approach is the targeting of contrast agents to folate receptors. These contrast media will identify pathologies with these receptors such as precancerous lesions or polyps. Unfortunately, the practical application of these agents continues to be limited by the fact that fairly high amounts are required to achieve an appreciable effect in MR imaging.

Simpler targeting mechanisms such as uptake of an agent into the lipid metabolism are used to identify atherosclerotic plaques. One such agent, the gadolinium derivative gadoflurine, enhances the signal intensity of plaques with a high lipid content.

Many studies have shown that iron oxide nanoparticles (USPIO) – whose use as blood pool and lymph node agents has already been discussed above – are also taken up by inflammatory cells (macrophages or histiocytes, lymphocytes) and may thus be used for so-called inflammatory imaging (with both T1- and T2-weighted sequences). Promising fields of application for this approach are the early identification of transplant rejection, demonstration of reactive atherosclerotic plaques, differentiation of acute from chronic glomerulonephritis, identification of inflammatory activity in certain types of multiple sclerosis plaques, uptake by synovial macrophages in rheumatoid arthritis, and peritumoral accumulation. In all of these cases, active endocytic cells can be identified by the extent of their signal changes.

In the monitoring of stem cell therapy, labeling of cells with USPIO has already been used with success.

12.3.8 Hyperpolarized Gases

The use of hyperpolarized gases as MR contrast agents is based on the polarization of nuclear spins. A high-volume production method is laser excitation of noble gases. The gases used for this purpose in medical imaging are helium-3 and xenon-129, which can be administered for evaluating lung ventilation or other hollow structures such as the gastrointestinal tract and the paranasal sinuses. The effects of hyperpolarized gases are visualized by using special pulse sequences with adjusted resonant frequencies and other optimization steps in order to achieve a high SNR. These optimized sequences yield diagnostic images of organs such as the lung, otherwise notoriously difficult to evaluate by MR imaging. The complexity of the equipment required, however, has so far hindered a wider use of this technique.

12.3.9 Oral MR Contrast Agents

Orally administered contrast agents facilitate the differentiation of a physiologic space from surrounding tissue and, at the same time, improve the distention of the intraluminal space. In the clinical setting, this mechanism is made use of in direct MR arthrography and to opacify the gastrointestinal tract. For MR arthrography, dilute solutions of unspecific Gd-based contrast agents (Artirem® 0.0025 mol/l, Magnevist® 2.0 with 0.002 mol/l) are injected directly into the joint space. The intracapsular fluid is thus clearly delineated from surrounding tissue by its high signal intensity on T1-weighted images and the joint space is markedly widened.

As with contrast agents used for computed tomography, gastrointestinal MR contrast media can be administered orally or rectally. In addition, butyl scopolamine (Buscopan®) or glucagon is given intravenously to reduce

Table 8. Examples of different types of gastrointestinal MR contrast agents

	Miscible contrast agents	Nonmiscible contrast agents
Positive contrast agents (SI increase)	Gd-DTPA (Magnevist® enteral) $MnCl_2$ = Lumenhance® Ferric ammonium citrate (Ferriselz®)	Fats Vegetable oils
Negative contrast agents (SI decrease)	Ferumoxsil (Lumirem®/Gastromark®) Barium sulfate Alumina	Perfluorocarbons CO_2

artifacts from peristalsis. Miscible agents that mix with the bowel contents and nonmiscible agents that do not are distinguished as well as positive and negative agents (▶ Table 8).

The positive water-soluble contrast media comprise unspecific Gd(III) complexes as well as iron and manganese solutions. The iron and manganese solutions undergo partial absorption. In the gastrointestinal formulation of Gd-DTPA (Magnevist® enteral), gadolinium is buffered with mannitol because it is much less stable in an acidic environment (buffer → less pronounced drop in pH). The buffer increases gastrointestinal osmolality, which results in further distention due to inflowing water and may cause diarrhea. The acid-fast Gd-DOTA has been studied in experimental investigations without addition of a buffer. Some juices rich in metal ions such as blueberry juice also increase the signal in the intestinal lumen.

The negative gastrointestinal contrast medium ferumoxsil (Lumirem®) consists of silicone-coated superparamagnetic iron oxide nanoparticles with many additional ingredients and is administered as a suspension. The side effects again include diarrhea. These agents are mainly used to suppress the gastrointestinal signal in MRCP and to help separate bowel loops from surrounding structures.

Barium sulfate and alumina decrease the signal by displacing water and hence the water protons. When higher doses are administered, the hypotonic effects of the suspension may cause obstipation. Perfluorocarbons (e.g. perfluorooctyl bromide) also decrease the signal by reducing the local proton density but have been abandoned due to their high price.

Water is probably the least expensive oral MR contrast medium which can be used to mark the intestinal lumen. It has a low signal on T1-weighted images and a high signal on T2-weighted images. Distention of the lumen can be improved by the admixture of gel formers or substances that increase osmolality (mannitol, PEG).

12.4 Outlook

The development of new contrast media for MR imaging makes high demands on both the pharmaceutical companies developing such agents and their partners in radiology because they must demonstrate a clinical benefit beyond simply that of improving image contrast and image quality. From a clinical perspective, the high diagnostic efficiency of modern imaging modalities is expected to lead to therapeutic advantages, indeed, more and more difficult to realize in today's ever changing technological environment. Modern contrast media should combine a maximum of clinical accuracy with an acceptable price, good tolerance, and easy handling in the routine clinical setting. These are the challenges facing the developers of new MR contrast media.

Table 9. Overview of clinically used MR contrast media and their most important properties

Brand name	Active ingredient	Concentration of complex	Element	Indication	Relaxivity in water at 1.0 T	Osmolality osm/kg H_2O	Thermodynamic stability at pH 7	Additional ingredients in formulation	Note
Artirem®	Same as Dotarem	Gd-DOTA 0.0025 mol/l	Gd^{3+}	Arthrography	See Dotarem	250-320			Dilute form of Dotarem®
Dotarem®	Gadoterate meglumine	Gd-DOTA 0.5 mol/l	Gd^{3+}	CNS, whole body, angiography	R1=3.4 R2=4.3	1350	$10^{18.8}$ $10^{25.8}$ (pH 9-10)		*Unspecific* water-soluble CM
Endorem® = Feridex™	Ferumoxides	11.2 mg Fe/ml	FeO	Focal liver lesions	R1=40 R2=160	340		Citric acid, mannitol	*Liver*: RES-specific CM dose: 0.075 ml/kg bw=15 µmol Fe/kg bw as infusion particle diameter: 160 nm contraindicated in hemosiderosis
Gadovist®	Gadobutrol	Gd-BT-DO3A 1.0 mol/l	Gd^{3+}	Perfusion, CNS, angiography	R1=3.6 R2=5.3	1603	$10^{15.6}$	Na-Ca butrol	*Unspecific* water-soluble CM double concentration: 50% reduction of volume or injection rate

Table 9. (continued)

Brand name	Active ingredient	Concentration of complex	Element	Indication	Relaxivity in water at 1.0 T	Osmolality osm/kg H₂O	Thermodynamic stability at pH 7	Additional ingredients in formulation	Note
Lumenhance®	Manganese chloride tetrahydrate	MnCl₂	Mn²⁺	Gastrointestinal tract				Glycine, polygalacturonic acid, sodium acetate, sodium benzoate, sodium bicarbonate, sugar, xanthane rubber, strawberry flavor	*Positive oral MR CM* lyophilisate 40 µg Mn²⁺/ml dose: 900 ml with 36 mg Mn²⁺ oral absorption (negligible amount of free Mn) approved in the US
Lumirem® = Gastromark™	Ferumoxsil	FeO 0.175 mg Fe/ml	FeO	Gastrointestinal tract oral, rectal		250		E110, E216, E218, ammonium glycyrrhizinate, sorbitol, saccharin Na, carboxymethyl cellulose	*Negative oral MR CM* 300 – 900 ml administered orally or rectally
Magnevist®	Gadopentetate dimeglumine	Gd-DTPA 0.5 mol/l	Gd³⁺	CNS, whole body; angiography	R1=3.4 R2=3.8	1940	$10^{17.7}$	0.2% dimeglumine-DTPA	*Unspecific* water-soluble CM
Magnevist® 2 mmol/l	Same as Magnevist	Gd-DTPA 0.002 mol/l	Gd³⁺	Arthrography		290			Lower concentration
Multihance®	Gadobenate dimeglumine	Gd-BOPTA 0.5 mol/l	Gd³⁺	Liver, CNS	R1=4.39 R1=9.7 (blood) R2=6.2	1970	10^{22} (pH 9-10)		*Liver:* hepatocyte-specific low *albumin* binding release of benzyl alcohol

Table 9. (continued)

Brand name	Active ingredient	Concentration of complex	Element	Indication	Relaxivity in water at 1.0 T	Osmolality osm/kg H_2O	Thermodynamic stability at pH 7	Additional ingredients in formulation	Note
Omniscan®	Gadodiamide	Gd-DTPA-BMA 0.5 mol/l	Gd^{3+}	CNS, whole body, angiography	R1=3.9 R2=5.1	790	$10^{14.9}$	5% CaNa-DTPA-BMA	*Unspecific* water-soluble CM
Optimark®	Gadoversetamide	Gd-DTPA-BMEA 0.5 mol/l	Gd^{3+}	CNS, liver		1110		0.05 mol/l calcium versetamide sodium, $CaCl_2$	*Unspecific* water-soluble CM approved for single-dose administration only (USA)
Primovist®	Gadoxetate disodium	Gd-EOB-DTPA 0.25 mol/l	Gd^{3+}	Liver	R1=4.7, R2=5.1 (1.5T, water) R1=7.4 (blood plasma)	690		Caloxetic acid, trisodium, trometamol	*Liver*: hepatocyte-specific 11% protein binding dose: 0.1 ml/kg bw caution in liver insufficiency
Prohance®	Gadoteridol	Gd-HP-DO3A 0.5 mol/l	Gd^{3+}	CNS, whole body, angiography	R1=3.7 R2=4.8 (0.47T)	630	$10^{17.1}$	0.1% Ca-HP-DO3A	*Unspecific* water-soluble CM
Resovist®	Ferucarbotran	28 mg Fe/ml	FeO	Focal liver lesions	R1=25.4 R2=151	333		Lactic acid, mannitol, NaOH	*Liver*: RES-specific CM dose: 0.9 ml (<60 kg), 1.4 ml (>60 kg) as a bolus particle diameter: 60 nm

Table 9. (continued)

Brand name	Active ingredient	Concentration of complex	Element	Indication	Relaxivity in water at 1.0 T	Osmolality osm/kg H_2O	Thermodynamic stability at pH 7	Additional ingredients in formulation	Note
Sinerem® = Combidex™	Ferumoxtran	210 mg Fe/g lyophilisate	FeO	(Lymph node staging)	R1=25 R2= 80-85 (0.47T)				*Lymph nodes* blood pool agent dose: 2.6 mg Fe/kg=45 µmol Fe/kg bw expected approval: 2007
Supravist®			FeO	Angiography, blood pool	R1=15.4 R2=42.9				*Blood pool agent* expected approval: 2008
Teslascan®	Mangafodipir trisodium	Mn-DPDP 0.01 mol/l (0.05 mol/l in the US)	Mn^{2+}	Liver	R1=2.3 R2=4.0	290		Vitamin C, NaCl	*Liver:* hepatocyte-specific CM manganese release and metabolism dose: 0.5 ml/kg bw=5 µmol/kg
Vasovist®	Gadofosvesete trisodium		Gd^{3+}	Angiography	R1=19 R2=37 (blood, 1.5T) R1=5.2 R2=5.9 (1.5T)				*Albumin binding* High protein binding expected approval in EU: 2006 (MS-325)

References

1. Bellin M-F, Webb JAW, Der Molen AJ, Thomsen HS, Morcos SK (2005): Safety of MR liver specific contrast media. Eur Radiol 15:1607–1614
2. Dawson P, Cosgrove DO, Grainger RG (eds) (1999): Textbook of Contrast Media. ISIS Medical Media, Oxford
3. Earls JP, Bluemke DA (1999): New MR imaging contrast agents. Magn Reson Imaging Clin North Am 7:255
4. Kirchin MA, Runge VM (2003): Contrast agents for magnetic resonance imaging: safety update. Top Magn Reson Imaging 14:403–425
5. Reimer P, Vosshenrich R (2004): Contrast agents in MRI. Substance, effects, pharmacology and validity. Radiologe 44: 273–283
6. Rohrer M, Bauer H, Mintorovitch J, Requardt M, Weinmann H-J (2005): Comparison of magnetic properties of MRI contrast media solutions at different magnetic field strengths. Investigative Radiology 40:715–724
7. Runge VM (2000): Safety of approved MR contrast media for intravenous injection. J Magn Reson Imag 2:205
8. Saeed M, Wendland MF, Higgins CB (1998): Blood pool MR contrast agents for cardiovascular imaging. J Magn Reson Imaging 12:890–898
9. Semelka RC, Helmberger TK (2001): Contrast agents for MR imaging of the liver. Radiology 218:27
10. Taupitz M, Schmitz S, Hamm B (2003): Superparamagnetic iron oxide particles: current state and future development. RÖFO 175:752–765
11. Webb JAW, Thomsen HS, Morcos SK (2005): The use of iodinated and gadolinium contrast media during pregnancy and lactation. Eur Radiol 15:1234–1240

13 MR Artifacts

13.1 Motion and Flow Artifacts (Ghosting)

Classical MR sequences are rather slow; it takes several minutes to acquire a T1-weighted image with an SE sequence. This is why MR imaging is highly sensitive to any kind of motion. Two types of motion artifacts are commonly encountered in routine MR imaging:
— Artifacts caused by breathing, peristalsis, or the beating heart (respiratory and cardiac motion artifacts)
— Artifacts caused by pulsatile blood flow or circulation of cerebrospinal fluid (CSF) (flow artifacts).

Motion Artifacts

Motion artifacts caused by respiration, the beating heart, and peristalsis are a common phenomenon in routine MR imaging and their occurrence was often used as an argument against performing abdominal MR imaging in the past. Motion artifacts degrade an MR image in the form of blurring or discrete ghosts. Ghosting caused by cardiac and respiratory motion is seen on chest MR images as noise running through the heart and mediastinum in the phase-encoding direction.

Various strategies have been devised to overcome such motion artifacts:
— The effects of respiratory motion can be minimized by means of specially developed compensatory algorithms (*respiratory compensation*). The simplest form is to acquire data only during maximum expiration (*respiratory gating*, analogous to *cardiac gating*). More sophisticated

techniques collect signals throughout the respiratory cycle, ordering the acquisition in such a way that the highest-quality data obtained during expiration fills the center of k-space where its contribution to image contrast is greatest.

— Use of fast GRE sequences for breath-held volume imaging. Alternatively, a series of slices may be acquired during repeated breath-holds. *Breath-hold imaging* yields better results than the use of respiratory compensation algorithms but takes longer and can only be performed in patients who are able to cooperate.

— Cardiac motion can be compensated for by synchronizing image acquisition with a specific phase of the cardiac cycle (*cardiac gating*). This is accomplished by simultaneous *electrocardiography* (*ECG*) *recording* and, for instance, triggering the RF excitation pulse to the R-wave of the ECG. In this case, TR is as long as one or several R-R intervals.

— Motion artifacts caused by peristalsis can be reduced by administration of a spasmolytic agent such as butyl scopolamine (*Buscopan*®).

— Parallel imaging (▶ Chapter 10) also helps reduce artifacts caused by cardiac motion, respiration, or peristalsis.

— The navigator technique can be used as an alternative to breath-hold imaging. This technique suppresses respiratory motion artifacts and can therefore be used, for instance, to image the heart with the patient breathing freely.

— A special phenomenon are CSF pulsation artifacts. These are intradural areas of low signal intensity that are predominantly seen on sagittal T2-weighted SE and FSE images. CSF pulsation artifacts can be prevented by using GRE sequences.

Flow Artifacts

Flow-related artifacts are caused by flowing blood as well as the flow of CSF and occur in the phase-encoding direction. These artifacts are due to the fact that spins moving along a magnetic field gradient (slice-selection, phase-encoding, or frequency-encoding gradient) experience a phase shift (see also phase-contrast angiography, ▶ Chapter 11.1.1). As a result, anatomy moving during the phase-sampling interval is assigned a wrong phase value and depicted in a different place in the image. Flow artifacts are typically seen as ghosts, i.e. structures that are not really present, such as a blood vessel which is depicted more than once in the phase-encoding direction.

The following remedies are available to prevent or reduce flow artifacts:

— *Flow compensation* or *gradient-moment nulling (GMN)*. Use of special gradient pulses applied prior to signal readout to compensate in advance for motion-induced dephasing.
— *Presaturation*. This is accomplished by saturating the blood on either side of the imaging slice immediately before actual data acquisition. The saturated blood does not produce any artifacts because it gives no signal when it enters the scan plane.
— Swapping the frequency- and phase-encoding axes may serve to remove an artifact that occurs only in the phase-encoding direction from the body region of interest.

Another remark on *presaturation*: The phenomenon of saturation as explained in ▶ Chapters 3 and 11.1.1 can be exploited to suppress specific tissue components by repeated excitation at short intervals. A tissue excited in this way emits no signal in the subsequent measurement because the spins have no time to recover between excitations. This technique is employed in time-of-flight angiography to suppress the signal from blood entering the slice from one direction while the blood entering from the other direction will continue to give a signal. In this way, one can selectively visualize either the veins or the arteries.

13.2 Phase Wrapping

Another notorious problem in clinical MR imaging is *aliasing* or *phase wrapping* (also known as *phase wraparound* or *foldover artifact*), which is caused by *phase encoding errors*. Wraparound artifacts occur whenever the dimensions of an object exceed the defined field of view. These parts are wrapped around and spatially mismapped to the opposite side of the image (▶ Fig. 56).

When a specific FOV is defined, the MR scanner assumes that the whole range of possible phase shifts from –180° through +180° occur within the FOV. Problems arise when the target anatomy extends beyond the FOV in the phase-encoding direction. In this case the parts outside the FOV are assigned a phase shift above +180° or below –180°. A phase of +190°, for example, corresponds to a phase of –170°. Objects with these phases are

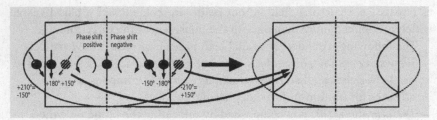

Fig. 56. Phase wrapping. Structures outside the defined field of view that are assigned the same phase shift as structures within the FOV are superimposed on the latter

assigned the same spatial encoding and therefore appear one on top of the other in the MR image. Structures extending beyond the right margin will be wrapped around to the left margin of the image and vice versa.

Various options are available to overcome the problem of wraparound artifacts:

A larger *field of view* can be defined to encompass all of the anatomy of interest. This will eliminate phase wraparound but at the cost of spatial resolution.

The frequency- and phase-encoding directions can be switched as there are no wraparound phenomena in the frequency-encoding direction (because the deep frequencies from one side are easily distinguished from the high frequencies from the other side). For example, when the chest or the pelvis is imaged, the gradients are applied so that the shorter (anteroposterior) dimension of the patient is oriented in the phase-encoding direction.

Special algorithms ("no phase wrap", "foldover suppression", or "antialiasing") prevent phase wrapping by oversampling in k-space: the field of view is enlarged so that no parts of the anatomy of interest extend beyond it. The excess data thus collected is discarded during image reconstruction. Note, however, that the no-phase-wrap option cannot be combined with some specialized imaging techniques.

Surface coils can be arranged in such a way that structures which otherwise might wrap around come to lie outside the sensitive range of the receive coil and thus do not appear in the image (e.g. in spinal imaging with an anteroposterior phase-encoding axis).

Presaturation (▶ Chapter 3.5) is another option to suppress the signal from regions outside the defined field of view.

13.3   Chemical Shift

The concept of chemical shift as introduced in ▶ Chapter 9 describes the fact that the resonant frequency of protons varies with their molecular environment. On MR scanners with a field strength of 1.0 T or greater, this phenomenon can be exploited to differentiate lesions with and without fatty components. In addition, chemical shift effects can be used to selectively suppress the signal from fat.

On the other hand, chemical shift phenomena also give rise to artifacts that are frequently encountered in medical MR imaging. Such chemical shift artifacts occur on the basis of two mechanisms: spatial misregistration between fat and water or silicone and water (chemical shift artifact of the first kind) and cancellation of the signal at the interface between fat and water (chemical shift artifact of the second kind).

Chemical Shift Artifact of the First Kind

Chemical shift artifacts of the first kind occur when protons with different precession frequencies (fat, water, and silicone) are depicted in a different place from where they are actually to be found along the frequency-encoding axis. This results from the fact that the signals from fat and water or the signals from silicone and water are spatially mismapped in the frequency-encoding direction. In medical imaging, chemical shift artifacts of the first kind occur at sites where fat and water are adjacent to each other or where fat is surrounded by water. Chemical shift misregistration manifests itself as a dark band (low or no signal) on the side of the higher spatial frequency and a bright band (high signal) on the side of the lower frequency (signal pile-up) (▶ Fig. 57). Bright signal bands are seen when protons with different resonant frequencies are depicted as if they coexist in the same voxel. Chemical shift artifacts of the first kind occur with all pulse sequences and their size depends on the receiver bandwidth and the magnetic field strength used. They can be reduced by broadening the receiver bandwidth. However, as we have seen in ▶ Chapter 5, a wider bandwidth will reduce SNR as well. Alternatively, chemical shift artifacts of the first kind can be reduced by swapping the frequency- and phase-encoding axes or by employing a fat suppression technique.

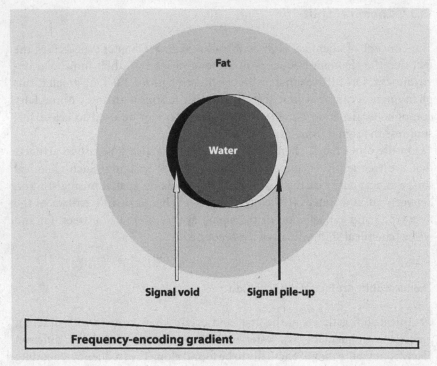

Fig. 57. Chemical shift artifact at a fat-water interface. Spatial misregistration of fat relative to water signal in the frequency-encoding direction results in a dark band (signal
void) on one side and a light band (signal pile-up) on the other. For details see text

Chemical Shift Artifact of the Second Kind

Chemical shift artifacts of the second kind are confined to GRE imaging.
Their characteristic appearance is a black rim (signal void) at the boundary
between fat and water. Such artifacts are seen, for example, at the interface
between perirenal fat and the renal parenchyma. They result from phase
cancellation effects when GRE images are acquired while fat and water are
out of phase. Chemical shift artifacts of the second kind can be avoided by
using SE sequences and can be minimized on GRE images by acquiring the
data with fat and water in phase.

13.4   Magnetic Susceptibility

Magnetic susceptibility is a fundamental property of all matter including biological tissues. It refers to the ability of a substance to become magnetized in an external magnetic field.

Metals typically have large susceptibilities. This property becomes relevant in medical MR imaging when patients with metal foreign bodies or implants are imaged. Such materials can lead to signal voids and/or image distortions at their boundaries with tissues that have different susceptibilities. This phenomenon is known as a susceptibility artifact. Less prominent susceptibility artifacts occur at tissue interfaces (e.g. between bone and muscle) or at interfaces between bone and air. An anatomic area especially prone to susceptibility artifacts is the transition between the paranasal sinus and the skull base. Other materials that may cause susceptibility artifacts are local deposits of calcium hydroxyapatite, accumulations of gadolinium chelate, or iron oxide particles.

In general, susceptibility artifacts may occur with all pulse sequences. They are minimal on SE images because the 180° refocusing RF pulse corrects for $T2^*$ effects and SE sequences themselves are fairly insensitive to static field inhomogeneities. On the other hand, the more pronounced susceptibility effects on GRE images can be exploited for diagnostic purposes, for instance to identify small hemorrhages or calcifications. In clinical MRI, minimizing susceptibility artifacts is especially important when body regions with orthopedic implants are imaged. Several strategies are available to reduce susceptibility artifacts from metal implants: use of SE and FSE sequences instead of GRE sequences, swapping of the phase- and frequency-encoding axes, imaging with a wider receiver bandwidth, alignment of the longitudinal axis of a metal implant with the axis of the main magnetic field, and use of STIR rather than frequency-selective fat suppression techniques.

13.5   Truncation Artifacts

Truncation artifacts are also termed ringing, Gibb's, or spectral leakage artifacts and arise as a consequence of using the Fourier transform to reconstruct an MR image. They typically appear as straight or semicircular parallel lines immediately adjacent to high-contrast interfaces such as the borders between muscle and fat or between CSF and the spinal cord. These artifacts are particularly problematic in spinal imaging, where they may mimic a syrinx or widening of the cord. Because truncation artifacts result from inadequate sampling of high spatial frequencies, they can be minimized by increasing the matrix in the phase-encoding direction.

13.6 Magic Angle

The magic angle artifact primarily affects structures with parallel fibers such as tendons and ligaments. These structures are characterized by a low signal intensity on most sequences because they have a short T2. Their signal may be increased and mimic pathology when the main magnetic field is at an angle of 55° to their fibers.

13.7 Eddy Currents

Eddy currents are generated when gradients are turned on and off quickly. These currents may be induced in the patient, in cables or wires around the patient, or in the magnet itself. Eddy currents generated by the magnet appear as a signal drop in the margin of the image. These artifacts can be reduced by optimizing the sequence of gradient pulses.

13.8 Partial Volume Artifacts

Partial volume artifacts occur whenever spatial resolution is limited. The signal intensities of different tissues and structures that are located in the same voxel are averaged. This may result in an intermediate signal at the interface between tissues with high and low signal intensities. The risk of partial volume artifacts can be reduced by increasing the number of slices acquired in the z-direction.

13.9 Inhomogeneous Fat Suppression

In the presence of a homogeneous magnetic field, uniform fat suppression (saturation) can be achieved by applying an RF pulse that has the resonance frequency of fat protons. However, in clinical imaging, this is rarely possible because the fat protons precess at different frequencies due to local field inhomogeneities, which may arise from the very presence of the patient within the magnet. Consequently, fat suppression is inhomogeneous because the RF pulse applied for fat suppression cannot match the different precessional frequencies of all fat protons.

Whenever considerable magnetic field inhomogeneities are likely to be encountered, for example, in patients with a metal foreign body, one should consider use of a STIR sequence for fat suppression because it will probably yield better results in such cases compared with presaturation in combination with SE, FSE, or GRE sequences.

13.10 Zipper Artifacts

A zipper artifact looks like a line of alternately bright and dark pixels running through the image in the phase-encoding or frequency-encoding direction. Zipper-like artifacts in the phase-encoding direction result from radiofrequency noise. Such noise may originate from an external source that reaches the receiver coil, for instance, because the door of the scanner room has not been fully closed. Another cause is RF emission from anesthesia monitoring equipment like pulse oximeters used within the scanner room. Zipper artifacts in the frequency-encoding direction are typically due to imperfect slice-selection profiles or inadequate RF transmission.

13.11 Crisscross or Herringbone Artifacts

A crisscross or herringbone artifact is due to a data processing or reconstruction error. It is characterized by an obliquely oriented stripe that is seen throughout the image. These artifacts can usually be eliminated by reconstructing the image again.

References

1. Peh WC, Chan JH (2001). Artifacts in musculoskeletal magnetic resonance imaging: identification and correction. Skeletal Radiol 30:179–191
2. Wood ML, Henkelmann WR (1999). Artifacts. In: Stark DD, Bradley WG (eds) Magnetic Resonance Imaging, 3rd edn. Mosby, St. Louis, p 215

14 High-Field Clinical MR Imaging

Dramatic advances have been made in medical MR imaging in recent years. The magnetic field strengths typically used for routine clinical imaging range from 0.2 to 1.5 T. Over the past several years, systems operating at higher field strengths have become more prevalent, particularly at research centers. At the same time, interest in clinical imaging at 3 T is increasing as well. Available data suggests that magnetic field strengths above 2 T involve no increased risk for patients. The maximum field strength approved by the US Food and Drug Administration (FDA) for routine clinical applications is 4 T. Current clinical interest focuses on 3-T scanners although it is already evident that even higher field strengths will be used to examine patients in the future. In the framework of scientific studies, MR scanners operating at 7 tesla have already been used in humans.

The 3-T MR imagers that are commercially available today do not differ from 1.5-T machines in terms of scanner architecture. These are whole-body scanners just like the MR systems operating at 1.5 T or lower field strengths.

The strongest argument in favor of switching to a higher field strength is the expected boost in signal-to-noise ratio (SNR) as the MR signal increases roughly in proportion to the field strength. In theory, the SNR would thus be doubled at 3 T compared with 1.5 T. The better SNR achieved with a high-field scanner can be used to improve spatial resolution or reduce imaging time. An improved spatial resolution might permit a better evaluation of anatomy so far only inadequately visualized by MRI. Alternatively, with shorter scan times, MR systems can be operated more economically because more patients can be examined. Finally, imaging at 3 T or even higher field strengths has the potential to improve more sophisticated applications of MRI such as functional imaging (spectroscopy or perfusion imaging and the like).

In conclusion, high-field MR imaging has both advantages and disadvantages which the user must be aware of when switching to this new technology.

14.1 Tissue Contrast

Higher field strengths alter the T1 and T2 relaxation times of biological tissues. T1 is usually longer at 3 T compared with 1.5 T while T2 is shorter. This means that one has to adjust the TRs and TEs of different pulse sequences when they are to be used on 3-T scanners. For spin echo and fast spin echo sequences, a longer TR is needed at 3 T to achieve a similar contrast as with 1.5 T. Conversely, TE should be somewhat shorter in order to compensate for the longer T1 relaxation times at 3 T.

14.2 Magnetic Susceptibility

Susceptibility effects (▶ Chapter 13.4) increase in proportion to the field strength of the magnet. As a result, image distortion may increase and degrade image quality especially when GRE sequences are used. Conversely, the stronger susceptibility effects may be advantageous in conjunction with MR techniques such as perfusion imaging (▶ Chapter 11.2) where they contribute to image contrast.

14.3 Chemical Shift

Chemical shift in Hz increases in proportion to the magnetic field strength. The larger chemical shift is advantageous in spectroscopy where the spectral lines are spread farther apart. This improves spectral resolution and discrimination of the peaks of fat and water, which in turn enables better calibration of the frequency-selective RF pulse for fat suppression. MR spectroscopy at 3 Tesla can be performed with a smaller scan volume, thereby reducing contamination of the spectrum from outside the area of interest.

14.4 Radiofrequency (RF) Absorption

The amount of energy deposited in the body by an RF field is proportional to the square of the field strength and is thus significantly greater for high-field scanners. The threshold for energy absorption in the body (primarily in the form of heat), defined as the specific absorption rate (SAR), is therefore more easily reached. This limits the scan times that are theoretically feasible on high-field scanners as the possible succession of pulses must be slowed down to prevent overheating. These limitations must be borne in mind when sequences optimized for 1.5 T are used on 3-T scanners. Specifically, four times as much RF energy needs to be applied per unit time to achieve the same flip angle at 3 T as at 1.5 T. A sequence with a pulse duration and amplitude optimized so that energy deposition is just below the SAR threshold at 1.5 T will exceed the upper limit at 3 T. This limits the use of sequences with high SAR values such as SE and FSE sequences.

Various strategies are available to minimize the overall SAR. A promising approach is to use a series of variable flip angles (VFA), which differ both in size and temporal spacing. The VFA strategy is associated with less energy exposure because the shorter intervals between two refocusing pulses reduce overall scan time while the resulting MR signal remains the same. Another promising technique is parallel imaging (▶ Chapter 10), which reduces RF energy deposition by applying fewer refocusing pulses per echo train while echo time is kept constant.

15 Bioeffects and Safety

The *static magnetic field* of an MR scanner may be extremely strong with field strengths in the range of 1.5–4.0 T (15,000–40,000 gauss). Such strong magnetic fields hold risks for both patients and personnel. A potential danger arises from ferromagnetic objects that may turn into dangerous missiles when brought near the magnet.

Most biomedical implants used today can be safely scanned at field strengths of up to 4.0 T. The metal components contained in many implants may induce artifacts on an MR image but the metals employed nowadays are typically not ferromagnetic and are therefore unlikely to get dislodged when exposed to the magnetic field of a medical MR scanner. This holds especially for the majority of orthopedic implants (including hip prostheses) which, along with most neurosurgical implants such as shunts, drains, tubes, or plates, no longer constitute contraindications to MR imaging. Caution is still advised in patients with cerebral aneurysm clips. Here, the MR-compatibility should be carefully established in each case although most clips used for treating skull base aneurysms today are MR-compatible. All clips placed to stop bleeding in peripheral vessels are safe. Most artificial cardiac valves implanted today are MR-compatible and yet again this should be established in each case.

Cardiac pacemakers are still contraindications to MRI because they contain a number of sensitive electronic components whose function may be impaired during the scan. Pacemaker electrodes are ideal antennas for the reception of RF energy, which may lead to arrhythmia. Moreover, the electrodes may heat up and thus cause burns or thrombosis of blood vessels. This also applies to most patients with transient pacemakers. In contrast, patients with a sternum cerclage can be imaged without problems. MRI follow-up of patients after coronary stenting should not be performed until at least six weeks after the intervention. Virtually all aortic and peripheral vascular stents currently used are also MR-compatible. Nevertheless, it is again recommended to check before an MRI examination.

At the time of printing, scanning is contraindicated in patients with internal defibrillators or left ventricular assist devices. Neurostimulators or cochlear implants are also considered contraindications.

In view of the possible risks just outlined, it is clear that a thorough history must be obtained prior to an MRI examination to carefully rule out any contraindications. Most centers use standardized questionnaires to elicit information about implants and other objects that might not be MR-compatible. In most cases, the questionnaire is supplemented by an oral interview.

Caution is also advised in patients with embedded metal fragments or bullets. As a general rule, the risk posed by these foreign bodies depends on their anatomic location and on whether they are ferromagnetic or not. Ferromagnetic fragments may be dangerous when found in a critical location such as the eye where they may damage the optic nerve if displaced during the scan. If the situation is unclear, an X-ray should be obtained prior to the MRI examination. Other critical locations of ferromagnetic objects are the brain, spine, lungs, mediastinum, and abdominal organs. Foreign bodies in other anatomic locations are usually safe to scan. Our policy is to monitor these patients more carefully and ask them to report any unusual sensations, especially while they are being moved into the bore of the magnet. Dental prostheses often contain ferromagnetic materials and we therefore ask patients to remove their prostheses, mainly because they may cause artifacts on MR images and not because of any risks they may pose.

Problems may also occur in patients with large tattoos, which may occasionally cause burns. Special monitoring is recommended. Piercings have also been reported to cause burns and should be removed before the scan.

Another safety issue is the exposure to the *changing magnetic fields* generated by the gradient coils. Concerns have been voiced that these fields may interfere with cardiac conduction and thus cause arrhythmia. Such an effect has not been observed with the gradient strengths used in routine clinical MR imaging at present. However, optical hallucinations like flashes of light have been attributed to these gradient fields and there is evidence that experimental MR imagers with much stronger and faster gradients stimulate peripheral nerves (especially when echo planar imaging is performed, ► Chapter 8.5).

There is an ongoing controversy about potential deleterious effects of changing magnetic fields on the fetus. It is known that cells in the phase of cell division (as during the first three months of fetal life) are sensitive to various physical effects. This is why MR imaging of a developing fetus should be delayed until after the first trimester.

Claustrophobia prevents a number of patients from undergoing an MR examination. Many more patients experience anxiety or are scared by the sheer bulk of the MR equipment. Whether an individual suffering from claustrophobia will be able to complete an MR examination crucially depends on whether the staff, while preparing the patient for the examination, can dissipate concerns through good care and detailed information about all aspects of the MR scan. Other measures that help patients tolerate the MR scan include drug sedation, mirrors placed within the scanner, or mirrored glasses to look outside. Today, scanning in an open-bore imager with a second vertical or horizontal opening is available as an additional alternative for imaging claustrophobic patients.

References

1. Hilfiker PR, Weishaupt D, Debatin JF (2002) Intravascular implants: Safety and artifacts. In: Arlart IP, Bongartz GM, Marchal G (eds) Magnetic resonance angiography, 2nd edn. Springer, Heidelberg Berlin, p 454
2. Edwards MB, Taylor KM, Shellock FG (2000) Prosthetic heart valves: Evaluation of magnetic field interactions, heating, and artifacts at 1.5 T. J Magn Reson Imag 12:363
3. Shellock FG (2003) Reference manual for magnetic resonance safety: 2003 edition. Amirsys, Salt Lake City
4. Elster AD, Link KA, Carr JJ (1994) Patient screening prior to MR imaging: A practical approach synthesized from protocols at 15 U.S. medical centers. Am J Radiol 162:195
5. Quirk ME, Letendre AJ, Ciottone RA, Lingley JF (1989) Anxiety in patients undergoing MR imaging. Radiology 170:463

Glossary

Arrows (→) refer to related entries elsewhere in the glossary.

3D Acquisition   Technique of volumetric imaging instead of acquisition of individual slices. Accomplished by performing → Phase encoding in two directions (phase-encoding and slice-selection gradients). *Advantages:* good → Signal-to-noise ratio, very thin slices can be obtained, excellent raw data set for secondary reconstruction, and → 3D MRA.

3D MRA   MR angiography based on 3D data acquisition. Typically, a volume is acquired during a single breath-hold. 3D MRA has become the standard MR technique for vascular imaging.

Acceleration factor   In parallel imaging, the factor by which the number of phase-encoding steps is reduced. The acceleration factor may range from 1.0 (no acceleration) to about 3.0-4.0.

Active shielding   Technique for containment of the fringe fields of an MR magnet. An actively shielded magnet consists of a set of two coils, an inner coil to generate the magnetic field and an outer coil to provide return paths for the magnetic field lines.

Aliasing   → Phase wrapping.

B₀   The static external magnetic field of an MR scanner. The field strength in clinical MR imaging ranges from 0.064–3.0 tesla (up to 8 T in experimental applications).

Black blood effect   Loss of the signal of flowing blood seen on spin echo images as a result of the fairly long echo times during which the excited blood leaves the scan plane and irreversible dephasing due to the different gradients.

Blips   The phase-encoding peaks in → Echo planar imaging.

Blood pool contrast agent   Higher-molecular-weight compounds or particulate agents with a long residence time in blood vessels, which results from the fact that their large molecular size prevents or slows down diffusion through the capillary walls. Also called intravascular contrast agents.

Blooming   Loss of signal observed at interfaces of calcium and tissue on GRE images. Blooming is a T2* effect.

Body coil   The integrated RF coil of an MR scanner.

Bound pool   → Bound protons.

Bound protons   Water protons not freely mobile in a tissue. They are macromolecular water protons bound by hydration. The incorporated water protons are restricted in their mobility and thus exchange less energy with their surroundings (long T1) while their fixed structure promotes their exchange with each other (extremely short T2 of < 0.1 msec). This is why bound protons do not contribute to the MR signal. → Free protons, → Magnetization transfer.

B-value   The b-value denotes how sensitive a sequence is to diffusion effects and thus represents a measure of the signal loss to be expected for a given diffusion constant. It is determined, among other things, by the strength and timing of the gradient pulses of the paired diffusion gradient and inversion pulse sandwich applied to make a sequence sensitive to diffusion effects.

Centric k-space ordering   Mode of data collection in which k-space is not filled in a linear fashion but from the center toward the periphery using a spiral trajectory (commercial implementations of this technique are CENTRA or elliptical centric ordering of k-space).

Chemical shift   Describes the fact that the resonant frequency of protons varies with their molecular environment. The chemical shift most important in clinical MR imaging is that between protons in fat and water. As a result of the chemical shift, the protons of fat and water which coexist in the same voxel may be alternately in phase, i.e. their transverse magnetization vectors add together, or out of phase (opposed phase), i.e. their magnetization vectors point in opposite directions. This phenomenon can be exploited to differentiate fatty tissue (signal drop on image acquired while fat and water are out of phase) from other tissue (no signal drop on out-of-phase image).

Chemical shift artifact   Spatial misregistration between fat and water signals in the frequency-encoding direction seen as a white or dark band at sites where fat and water are adjacent to each other (chemical shift artifact of the first kind). Chemical shift artifacts of the second kind designate the signal losses resulting from phase cancellation effects on GRE images obtained with fat and water out of phase.

Coil   Component of an MR scanner which serves to transmit RF pulses and/or receive MR signals.

Coil array   Arrangement of several surface coils placed side by side for simultaneous signal collection in parallel imaging.

Contrast-to-noise ratio (CNR)   Measure of the ability to differentiate two adjacent anatomic structures in an MR image on the basis of their signal intensities in relation to image noise.

Crisscross artifact   Also called herringbone artifact. Artifact caused by a data processing or reconstruction error. Can usually be eliminated by reconstructing the image again.

Cross-talk   Interference resulting from the unintended excitation of adjacent slices which overlap at their edges due to imperfect, nonrectangular slice profiles. Cross-talk decreases → Signal-to-noise ratio.

Dixon technique   MR imaging technique for the reconstruction of fat and water images based on the → Chemical shift between fat and water.

ECG gating   MR imaging technique that acquires data only during a specific phase of each cardiac cycle (e.g. systole or diastole) (→ Gating).

Echo planar imaging (EPI)   A gradient echo technique that uses an ultra-fast → Frequency-encoding gradient to generate a series (train) of up to 128 gradient echoes. EPI thus enables single-shot acquisition of an image in less than 100 msec.

Echo time (TE)   The interval between excitation of a spin system and collection of the MR signal. TE predominantly determines the amount of T2 contrast of the resultant image.

Echo train length (ETL)   Number of echoes sampled per TR when a → Fast spin echo sequence is used.

Eddy currents   Electrical currents induced when the gradients turn on and off. These currents cause a drop of signal in the margin of the MR image.

Effective echo time   In an FSE sequence, the time between the excitation pulse and the echo which primarily determines T2 contrast because it produces the strongest signal.

EPI   → Echo planar imaging.

Ernst angle   The → Flip angle at which the maximum signal is generated for a given TR and TE.

Excitation angle   → Flip angle.

Exorcist   Compensatory algorithm which is applied to reduce → Ghosting caused by breathing, hence the name Exorcist.

Extracellular contrast agent   Low-molecular-weight, water-soluble compound with distribution in the vascular and interstitial spaces of the body after IV administration. Most of the MR contrast agents in clinical use today belong to this group of gadolinium (III) complexes.

Fast spin echo sequence (FSE)   A → Spin echo sequence run more rapidly than usual; also known as turbo spin echo or RARE. This technique shortens scan time by generating up to 16 echoes with a series of 180° pulses. FSE sequences have the same image quality as conventional SE sequences and are nearly as fast as GRE sequences.

Fat saturation   (Fat sat, fat suppression) Various techniques are available to eliminate the signal from fatty tissue. One fat suppression technique employs an RF pulse which is shifted by 220 Hz (at 1.5 T) and thus selectively saturates fat protons (frequency-selective fat saturation). Alternatively, fat suppression can be achieved by making use of the → Chemical shift between fat and water or by using a → STIR sequence.

Ferromagnetism   Property of a material, as iron, of becoming permanently magnetized. Ferromagnetic materials can markedly distort the magnetic field and cause large signal voids in the MR image.

FFE   Fast-field echo → Gradient echo sequence.

Field of view (FOV)   The area of anatomy covered in an image. The FOV is usually square, though a → Rectangular FOV may be chosen to reduce scan time. A smaller FOV improves spatial resolution but decreases → Signal-to-noise ratio.

FLAIR   (Fluid-attenuated inversion recovery) Variant of an inversion recovery sequence which is based on a fast spin echo sequence and uses a very long inversion time. This sequence is primarily used in neuroradiologic imaging because it completely suppresses the signal from cerebrospinal fluid and thus improves the detection of lesion that are otherwise difficult to differentiate from surrounding brain tissue.

Flip angle   (Excitation angle, pulse angle). The angle by which magnetization is tilted when a spin system is excited by an RF pulse. The angle can be varied freely by changing the strength and duration of the excitation pulse applied. A flip angle of exactly 90° deflects all longitudinal magnetization (M_z) into the transverse plane (xy-plane). The flip angle is always 90° in a → Spin echo sequence while a → Gradient echo sequence can be acquired with different flip angles, e.g. 30°. The flip angle determines the amount of T1 weighting of an MR image.

Fourier transform   Mathematical operation needed to reconstruct MR images from raw data. The Fourier transform decomposes the measured MR signal into its frequency spectrum. In medical MR imaging, two-dimensional and three-dimensional Fourier transforms (2D-FT, 3D-FT) are used for image reconstruction.

FOV   → Field of view.

Fractional echo imaging Technique used to reduce scan time. Only half (or slightly more than half) the lines of k-space in the frequency-encoding direction are filled. Also known as partial echo imaging. → Partial k-space acquisition.

Free induction decay (FID) Signal loss that occurs at a characteristic time constant T2* without any external influence.

Free protons The free protons (protons in free water) of a tissue interact frequently with their environment (short T1) but rarely with each other (long T2). Only free protons contribute to the MR signal. → Bound protons, → Magnetization transfer.

Frequency encoding Part of → Spatial encoding of an MR signal. While the echo is being sampled, a gradient field is switched on in one dimension, imparting different precessional frequencies to the nuclear spins along that dimension. In this way, a spectrum of resonance frequencies is obtained instead of a single frequency (→ Fourier transform). The frequency information serves to locate the individual signal components in space along the gradient.

Frequency-encoding gradient The gradient field that is switched on while the MR signal is being collected, hence it is also called readout gradient. It is needed for → Frequency encoding of the MR signal.

FSE → Fast spin echo sequence.

Gating Technique of synchronizing MR imaging with the respiratory or cardiac cycle. ECG gating serves to reduce artifacts caused by cardiac motion. This is accomplished by triggering the scan to the R-wave of the ECG, thereby collecting data from the same phase of the cardiac cycle with each acquisition.

Ghosting Misencoding resulting in noise running through the heart and mediastinum or the multiplication of an anatomic structure such as the aorta in the phase-encoding direction. These artifacts are typically caused by pulsatile flow, less frequently by the beating heart or breathing.

Gibb's artifact → Truncation artifact.

Gradient   Defines the strength of the change of a quantity in a specific spatial direction. A magnetic field gradient in MR imaging refers to the linear change in magnetic field strength created along the x-, y-, or z-axis of the stationary magnetic field. Such gradients are needed for slice selection (→ Slice-selection gradient) and → Spatial encoding and are generated using dedicated coils built into the scanner. In a more general sense, the term "gradients" is also used to denote the gradient coils.

Gradient echo sequence (GRE)   Pulse sequence which differs from a → Spin echo sequence in that no 180° refocusing pulse is applied. Magnetic field inhomogeneities and the phase differences imparted by the gradient are not compensated for and the MR signal decays with T2* instead of T2. *Advantage:* shorter scan time.

GRASE   (Gradient and spin echo) A hybrid pulse sequence that combines a → Fast spin echo sequence and → Echo planar imaging. Several spin echoes are generated and, for each SE, several gradient echoes are acquired. *Advantages:* short scan time and high contrast (as with → Spin echo sequence). *Disadvantages:* technically demanding; clinical role still unclear.

GRE   → Gradient echo sequence.

Hyperpolarized gases   MR contrast agents for special indications. They are produced by laser polarization of the nuclear spins of noble gases (e.g. helium-3, xenon-129).

Inflow angiography   → Time-of-flight angiography.

Inflow effect   (Flow-related enhancement) Describes the fact that fast → Gradient echo sequences depict blood flowing into the scan slice with a bright signal while stationary tissue appears dark due to → Saturation.

In phase   → Chemical shift.

Intermediate-weighted image   → Proton density-weighted image.

Interslice gap   The distance between the nearest edges of two adjacent slices.

Intravascular contrast agent → Blood pool contrast agent.

Inversion recovery sequence (IR sequence) Spin echo sequence with an additional 180° inversion pulse preceding the usual excitation and refocusing pulses (→ Inversion time). Two IR sequences widely used in clinical MR imaging are → STIR and → FLAIR.

Inversion time (TI) The interval between the 180° inversion pulse and the 90° excitation pulse in an → Inversion recovery sequence. The TI can be selected to null the signal from a specific tissue such as fat, which is done by applying the 90° RF pulse when the magnetization of that tissue is zero.

IR Inversion recovery (→ Inversion recovery sequence).

Isocenter The geometric center of the main magnetic field of an MR scanner where the field strength is not affected by any of the three gradients.

K-space The mathematical space for storage of the measured raw data before the MR image is reconstructed by applying 2D or 3D → Fourier transform. The center lines of k-space predominantly determine image contrast while the peripheral lines mainly affect spatial resolution.

Larmor frequency Frequency at which spins precess about a magnetic field. The precession or resonance frequency is proportional to the strength of the magnetic field applied.

Longitudinal relaxation → T1 relaxation.

Magnetic susceptibility Measure of the extent to which a tissue or substance becomes magnetized when placed in an external magnetic field.

Magnetization transfer Describes the transfer of magnetic saturation from bound macromolecular protons to free protons. This phenomenon reduces the signal intensity of free water.

Matrix Two-dimensional grid consisting of rows and columns in which each square is a pixel (picture element). The matrix determines the number of pixels that make up an image.

MIP (Maximum intensity projection) Technique of image reconstruction which filters out the high signal intensities and projects them onto a single plane.

MR angiography MR technique that uses sequences providing good vessel-tissue contrast for generating MR angiograms. → Phase-contrast angiography, → Time-of-flight angiography, → 3D MRA.

MR arthrography MR technique for imaging of the joints, usually performed with intra-articular administration of a dilute gadolinium chelate solution under fluoroscopic guidance. The contrast medium widens the joint space, thereby improving the evaluation of intra-articular structures and hence the diagnosis of certain joint disorders.

Navigator MR technique for the suppression of respiratory motion artifacts which uses additional echoes (navigator echoes) to detect changes in the position of the diaphragm. The MR images are then reconstructed using only the data acquired with the diaphragm in a specific position. Using the navigator technique, it is possible to perform cardiac imaging with the patient breathing freely.

Negative contrast agent MR contrast agent that improves contrast by causing a selective signal loss in specific tissues accumulating the agent. Negative agents usually contain paramagnetic or superparamagnetic substances. → Paramagnetism, → Superparamagnetism.

NEX, NSA (Number of excitations, number of signal averages) Denotes how often a signal from a given slice is measured per phase encoding. An increase in NEX usually improves → Signal-to-noise ratio.

Opposed phase → Chemical shift.

Out of phase → Chemical shift.

Outflow effect → Black blood effect.

Parallel imaging Fast MR imaging technique with simultaneous signal collection by means of several surface coils placed side by side.

Paramagnetism A property exhibited by substances which are magnetized when exposed to an external magnetic field, resulting in a local increase in the magnetic field. A typical paramagnetic substance is the metal ion Gd^{3+}, which is used as an MR contrast medium in its chelated form. When low concentrations are administered, this compound shortens T1 and thus acts as a → Positive contrast agent. At higher concentrations, gadolinium complexes cause a signal loss due to local magnetic field inhomogeneities. → Magnetic susceptibility, → Negative contrast agent.

Partial Fourier imaging Technique of k-space filling in which only slightly more than half the k-space lines in the phase-encoding direction are actually sampled and the unfilled lines are interpolated. The scan time is thus reduced by almost 50% while resolution is the same but noise is somewhat increased. → Partial k-space acquisition.

Partial k-space acquisition General term for different techniques employed to reduce scan time by incomplete sampling of the lines of k-space: → Rectangular FOV, → Partial Fourier imaging, → Fractional echo imaging.

Partial volume effect The loss of contrast between two adjacent tissues with different signal intensities caused by insufficient resolution when both tissues are in the same voxel.

Phase The angle by which a rotating magnetic vector of a precessing spin in the xy-plane differs from that of a second vector.

Phase-contrast angiography Technique that applies an additional gradient to encode the velocity of flowing spins (e.g. in flowing blood). Phase-contrast angiography is an → MR angiography technique that allows precise measurement of blood flow velocity.
Advantages: sequence can be sensitized to different flow velocities by user; technique allows quantitative determination of flow velocity.
Disadvantages: long scan time due to additional gradients and separate measurement for each direction to which the sequence is sensitized; pulsatile flow causes artifacts.

Phased-array coils An arrangement of coils consisting of several surface coils used simultaneously to improve image quality. Such an array combines the signal of a surface coil with the FOV of a body coil and enables the acquisition of high-resolution images of organs deep within the body such as the pelvic organs.

Phase encoding   Part of → Spatial encoding. Accomplished by switching a gradient to impart different phase shifts to the spins in an excited slice according to their position along the gradient. Spatial position can then be identified by a unique amount of phase shift.

Phase-encoding gradient   The gradient that is switched on for → Phase encoding during readout of the MR signal.

Phase wrapping   Phenomenon which occurs when parts of the anatomy of interest extending beyond the defined field of view are wrapped around and spatially mismapped to the opposite side of the image.

Pixel   Two-dimensional picture elements which make up the → Matrix.

Positive contrast agent   A positive MR contrast agent improves contrast by enhancing the signal, thereby making the tissue appear bright. Most positive MR contrast agents shorten T1.

Prepulse   → Presaturation.

Presaturation   Selective magnetic saturation of a tissue by applying an extra RF pulse (prepulse) immediately before the excitation pulse for generating the signal is delivered. Presaturation is performed to eliminate artifacts or to selectively suppress the blood signal (outside the scan plane) and to increase T1 weighting (within the scan plane).

Proton density-weighted image   Proton density-weighted (PD images), density-weighted, or intermediate-weighted MR images are images whose contrast is predominantly determined by the proton density of the tissues imaged. They are acquired with a fairly long repetition time (to minimize T1 effects) and a fairly short echo time (to minimize T2 effects). PD images have a high → Signal-to-noise ratio. A typical parameter combination for obtaining a PD image is TR/TE=2000/20 msec.

Quench   Sudden loss of superconductivity with breakdown of the magnetic field.

R1 and R2   Relaxivities: R1=1/T1 and R2=1/T2, unit: $(\text{sec mol/l})^{-1}$

Readout   The sampling of the MR signal.

Readout gradient   → Frequency-encoding gradient.

Receiver bandwidth   The spectrum of spin frequencies registered in MR imaging during readout.

Rectangular FOV   Technique of → Partial k-space acquisition with sampling of fewer k-space lines in the phase-encoding direction. A rectangular field of view is used to reduce scan time compared with full acquisition and is achieved at the cost of slightly reduced → Signal-to-noise ratio.

Region of interest (ROI)   Refers to a small area in a tissue that is selected, for example to measure signal intensity.

Relaxivity   Denotes the ability of a substance to change the relaxation time of a tissue; mainly used to describe the effect of an MR contrast agent on T1 (R1) and T2 (R2). It is usually given as molar relaxivity and varies with temperature and field strength.

Repetition time (TR)   The interval between two successive excitations of the same slice. By changing the TR, the user can determine the amount of T1 contrast of the resultant image.

Resonance frequency   Frequency at which resonance occurs, corresponds to the Larmor frequency of protons.

Respiratory compensation   (Resp comp) Algorithm which reduces artifacts due to respiratory motion by synchronizing scanning with the respiratory cycle. Also known as → Exorcist.

Respiratory gating   Scanning during a specific phase of the respiratory cycle (e.g. during inspiration or expiration). Typically performed using a respiratory belt to monitor the respiratory rate.

Ringing artifact   → Truncation artifact.

Rise time Parameter that describes the performance of a gradient. It is the time it takes to reach maximum gradient amplitude.

SAR (Specific absorption rate) Measure of the amount of energy deposited by an RF pulse in a certain mass of tissue. The energy applied during an MR experiment leads to tissue heating, which must not exceed certain thresholds defined in official guidelines.

Saturation Magnetic saturation causes a signal loss when → Repetition time is very short because there is not enough time for complete recovery of magnetization between two excitations. This can be remedied by reducing the → Flip angle. → Gradient echo sequence.

Scan time Also known as image acquisition time. Scan time is the key to the economic efficiency of an MR scanner and is determined by the number of phase-encoding steps, number of excitations (→ NEX), → Repetition time, and → Echo train length.

Shimming Correction of magnetic field inhomogeneities.

Signal-to-noise ratio (SNR) Measure of image quality expressed as the relationship between signal intensity and image noise.

Slew rate Parameter that describes the performance of a gradient, defined as the maximum gradient amplitude divided by the → Rise Time.

Slice-selection gradient Data collection requires the selective excitation of a slice, which is done by applying a slice-selection gradient.

SNR → Signal-to-noise ratio.

Spatial encoding All measures needed to determine the spatial origins of the different components of an MR signal. Spatial encoding comprises → Phase encoding and → Frequency encoding.

Spin Fundamental property of almost all elementary particles (protons, neutrons, and electrons). Spin denotes the magnetic properties that result from the angular momentum of a particle and hence relates to its ability to undergo nuclear magnetic resonance. In theory, all atomic nuclei with spin could be used for MR imaging (e.g. phosporus or fluorine) but hydrogen nuclei, which consist of a single proton, are used for clinical MR imaging because of their abundance in biological tissues.

Spin echo sequence (SE)   Most widely used pulse sequence in routine clinical MR imaging. It consists of an excitation pulse with a flip angle of exactly 90° which is followed by a 180° RF pulse for refocusing the spins after dephasing caused by T2* effects has occurred. It is a robust sequence that is insensitive to magnetic field and gradient inhomogeneities but is limited by a long scan time.

SPIO   (Superparamagnetic iron oxide particles) Iron oxide nanoparticles that are mainly used as RES-specific contrast media in liver imaging. SPIO particles have a larger diameter than → USPIO.

SPIR   (Spectral presaturation with inversion recovery) Strictly speaking, SPIR is not a complete MR sequence but merely a 180° prepulse which is made frequency-selective and only inverts fat magnetization. It can be combined with other sequences to acquire fat-saturated images.

Spoiling   Technique of spin dephasing which is mostly employed in conjunction with GRE imaging. A spoiled GRE sequence is a pulse sequence in which a spoiler gradient or RF spoiling is applied to destroy transverse magnetization before the next excitation pulse is applied. Spoiled GRE sequences are used to produce T1- or T2*-weighted images.

SSFP   (Steady-state free precession) GRE technique in which longitudinal and transverse magnetization contribute to the MR signal and contrast is determined by the relationship between T1 and T2. Examples of SSFP sequences are true FISP, FIESTA, and balanced FFE.

STIR   (Short TI inversion recovery) An → Inversion recovery sequence used to suppress the signal from fat, which is accomplished by selecting the inversion time such that the 90° RF pulse is applied when fat magnetization passes through zero. This technique suppresses all signals from tissues with short T1 values similar to those of fat.

Superparamagnetism Greatly increased → Paramagnetism (10- to 1000-fold). An example of superparamagnetic substances used as MR contrast agents are iron oxide nanoparticles. They can serve as a → Negative contrast agent.

Susceptibility artifact Signal loss due to the magnetic susceptibility of a tissue or other material.

T1 Tissue-specific time constant of → T1 relaxation which depends on the magnetic field strength, B_0, and is in the range of one to several seconds at 1.5 T.

T1 relaxation Also called spin-lattice relaxation and longitudinal relaxation. It refers to the return of excited spins to the equilibrium state or recovery of longitudinal magnetization and is associated with the transfer of energy to the surroundings.

T1-weighted image (T1w) MR image whose contrast is mainly (but not only!) determined by T1. T1 weighting is achieved by combining a rather short repetition time with a short echo time (to minimize T2 effects). Example: TR/TE=500/20 msec. Tissues with a short T1 appear bright while tissues with a long T1 appear dark.

T2 Tissue-specific time constant of → T2 relaxation. It is in the range of up to several hundred milliseconds and is independent of the magnetic field strength.

T2 relaxation Also called spin-spin relaxation and transverse relaxation. Dephasing of spins resulting from spin-spin interaction and energy exchange with each other. There is no energy transfer to the surroundings.

T2-weighted image (T2w) MR image whose contrast depends primarily on T2. T2 weighting is achieved by combining a long repetition time (to minimize T1 effects) with a long echo time. Example: TR/TE=2000/80 msec. Tissues with a long TR are bright on T2-weighted images while tissues with a short TR are dark..

T2* Time constant of → T2* relaxation.

T2* contrast   Image contrast that results from the specific T2* decay constants of different biological tissues. The T2* contrast of a GRE image can be manipulated by changing the echo time (TE).

T2* relaxation   All processes that contribute to spin dephasing. T2* relaxation comprises pure spin-spin interaction (→ T2 relaxation) and the effects of static magnetic field inhomogeneities. Application of a 180° RF pulse cannot reverse T2 relaxation itself but only the loss of phase coherence due to static field inhomogeneities. → Spin echo sequence.

TI   → Inversion time.

Time-of-flight (TOF) angiography   (Inflow angiography) An MR angiography technique that is based on the → Inflow effect. Well suited for imaging of the veins while arterial TOF angiography is impaired by artifacts. Contrast-enhanced → 3D MRA is the perferred option for imaging of the arteries.

Time-resolved MRA   The term time-resolved MR angiography is now mostly used to designate the dynamic study of the distribution of a contrast medium in the vascular system. Such a dynamic study is performed by rapidly and repeatedly imaging a vascular region following administration of a single dose of a contrast agent. The individual MRA images obtained in this way represent different phases of progressive contrast medium distribution.

TIRM   (Turbo inversion recovery magnitude) → FLAIR.

Transverse relaxation   → T2 relaxation.

True FISP sequence   → Gradient echo sequence in which the signal intensity in the steady state is determined by the T2/T1 ratio.

Truncation Artifact   (Gibb's artifact, spectral leakage artifact) Truncation artifacts are bright or dark lines that are seen parallel or adjacent to borders of abrupt intensity changes, as for example at the border between the bright CSF and the dark spinal cord on T2-weighted images. In the spinal cord, this artifact can simulate a syrinx. It is also noted in other locations of the brain/calvarium interface. This artifact is related to the finite encoding steps used by the Fourier transform to reconstruct an image.

TSE (Turbo spin echo) → Fast spin echo sequence.

USPIO (Ultrasmall superparamagnetic iron oxide particles) Very small iron oxide nanoparticles mainly used as lymph node-specific MR contrast agents.

Voxel Volume element that is represented by a → Pixel in the two-dimensional MR image; voxel size determines → Signal-to-noise ratio and spatial resolution.

Zero filling Technique of incomplete k-space filling. The portions of k-space that are not directly sampled are filled with zeros. In this way, a larger matrix is reconstructed by interpolation. Zero filling techniques are mainly used to reconstruct images in MR angiography.

Subject Index

168